VECTOR
BACKGROUND

ESSENTIAL OFF-GRID SURVIVAL PROJECTS

ESSENTIAL OFF-GRID SURVIVAL PROJECTS

Published by Cider Mill Press, an imprint of HarperCollins Focus LLC, 501 Nelson Place, Nashville, TN 37214 USA.

13-Digit ISBN: 978-1-40035-548-8

10-Digit ISBN: 1-40035-548-6

Books published by Cider Mill Press Book Publishers are available at special discounts for bulk purchases in the United States by corporations, institutions, and other organizations. For more information, please contact the publisher.

cidermillpress.com

HarperCollins Publishers,
Macken House, 39/40 Mayor Street
Upper, Dublin 1, D01 C9W8,
Ireland (https://www.harpercollins.com)

Typography: Text is set in Macklin Slab and TT Norms. Headers are set in Cralter Sans and Belfort.

Image Credits: Projects and part openers illustrated by Roelien Immelman. Part opener backgrounds and footer art licensed from Adobe Stock (pp. 16, 110, 154, 206).

Interior and cover design by Will Mack.

Printed in Malaysia

26 27 28 29 30 PJM 5 4 3 2 1

First Edition

ESSENTIAL OFF-GRID SURVIVAL PROJECTS

50 LIFE-SAVING SOLUTIONS

for SHELTER, WATER, FOOD and SAFETY

KY FURNEAUX

CIDER MILL PRESS
BOOK PUBLISHERS

CONTENTS

INTRODUCTION

WHEN DID PREPPING TURN INTO A DIRTY WORD?

The term **prepping** has, unfortunately, become something of a punchline in popular culture. Much of this stems from the way the media has sensationalized survivalists—often portraying extreme characters hoarding supplies, building elaborate bunkers, or obsessing over apocalyptic scenarios. These portrayals tend to focus on the fringes, creating caricatures rather than offering a fair and nuanced understanding of what preparedness actually involves. As a result, prepping is often dismissed as a fringe hobby rather than a practical approach to resilience.

Skepticism about the likelihood of major disasters has further contributed to the idea that prepping is an overreaction. This makes it easy for critics to write it off as eccentric or paranoid rather than a thoughtful response to uncertainty.

Another common misconception is that prepping stems from fear. In truth, it's the opposite. Being prepared helps reduce fear and uncertainty because it offers a sense of security—knowing that even if the worst happens, you've got it covered. It's about peace of mind, not pessimism.

There's a saying that sums it up well: "Expect the best but prepare for the worst."

For me, preparing allows space for optimism. And the best part? You don't need to be a survival expert to get started.

My journey into the world of off-grid living and survival skills has taught me that preparation isn't just for the "survivalists" or the rugged outdoorsy types; it's for anyone willing to learn and adapt. Whether you're a seasoned expert or someone just starting, this book contains easy-to-follow, step-by-step instructions on how to turn any home into a life raft.

Why should you prepare? Because anyone can learn basic off-grid living skills, and those skills can become invaluable in times of crisis.

This book is structured in a project-based format, meaning each chapter is a clear, manageable step toward building your resilience. You'll find consistent templates for each project, detailing when and how to do it, and the required tools. Whether you're looking to enhance your food storage, secure a water supply, ensure safety, or create shelter, you can jump directly to the sections that address your specific needs or follow the entire program for a comprehensive approach.

Imagine a world where disaster strikes: The lights flicker and then fade, the hum of appliances ceases, and the familiar sounds of a bustling life fall silent. In such moments, those who have cultivated a preparedness mindset stand apart. They have not only gathered essential supplies but have also equipped themselves with the skills and knowledge to thrive in adversity.

But preparedness is more than just gathering supplies; it's about cultivating a mindset that embraces flexibility, safety, and collaboration. I encourage you to involve your spouse, children, and even neighbors in these projects. Turn them into opportunities for skill-building and confidence-enhancing activities. Create family drills, practice plans regularly, and make preparedness a part of your family culture. By working together, we not only prepare ourselves for emergencies but also strengthen our bonds and build a support network. This mindset encourages a proactive response rather than a reactive one.

In a world increasingly reliant on technology and the conveniences of modern living, the notion of a preparedness mindset can often feel distant or even unnecessary. However, a recent increase in natural disasters as well as the constant threat of potential hacks to technology transforms that mindset from a mere concept into a critical lifeline. As someone who has embraced the principles of survival and self-sufficiency, I can attest to the profound importance of being ready for unexpected challenges. A preparedness mindset fosters resilience, enabling us to navigate the tumultuous waters of uncertainty with confidence and capability.

All of that said, this book is designed for the everyday person who, in the back of their mind, has a feeling that there is something more they can do to prepare for the unpredictable nature of the world we now live in.

The projects within these pages are family-friendly, affordable, and adaptable to various situations, including accommodating pets, medical needs, or limited living spaces. But remember: The time to act is now—**before** an emergency occurs. Together, let's embark on this journey toward preparedness, making it a fulfilling and enriching part of our lives.

THE BENEFITS OF A PREPAREDNESS MINDSET

SURVIVAL, WHETHER WILDERNESS OR URBAN, triggers the same emotional and physiological states within us. For this reason I think it's important to begin by discussing what happens in survival states and how best to combat them mentally.

Survival has baffled people for centuries. Why, in the same given circumstances, does one person survive when another doesn't? Phrases such as "the will to live" are often thrown around, but when you break them down, they seem to suggest that the mind holds such powerful influence over the body that it can simply decide to give up, causing vital bodily functions to shut down. This defies logic, yet time and again survivors of emergency scenarios have described the light fading in someone's eyes as they give up when other people in the same situation push through.

There have been real-life cases where people had access to food, water, and shelter yet still perished, often because the psychological weight of their situation became too much to bear. Some believe these individuals die because they became depressed and simply gave up, unable to adapt to a radically altered way of life.

To better understand the mental aspects of an unexpected emergency, we need to look at what actually happens when such a crisis strikes. From there, we can begin to identify tools and strategies to protect our bodies and minds.

As soon as an unexpected scenario occurs, the body's fight-or-flight response is triggered, releasing a surge of adrenaline. This adrenaline can boost your energy enough to escape a life-or-death moment, but as that moment extends into minutes/hours/days, it can actually inhibit your ability to make good decisions. The brain can process only so much information at once. When overloaded with anxiety, its cognitive functions can become restricted. This can lead to an inability to think or make decisions, where a person becomes mentally "frozen" and unable to respond effectively in a dangerous situation.

It has been recorded that during trauma, people regularly experience:

- Cognitive paralysis (complete inaction)
- Perseveration (repeating a particular response regardless of what is happening)
- Stereotypical behavior (repeating a behavior that might not be suitable to the situation without being able to change it)
- Hyperactivity (constant activity with a tendency to impulsive and aggressive behaviors)
- Hypoactivity (the inability to motivate yourself to action)

These responses can cause people to become disorganized, act inappropriately for the circumstances, or react too slowly to changing conditions.

Survival demands the ability to cope with environmental stress and requires the flexibility to adapt to that environment.

Some behaviors that I have identified as contributing to survival success include:

- Being prepared: Having clear, actionable directives and equipment available to ensure your basic needs are covered.
- Goal-directed behavior: Making a list of tasks you can complete and then working to complete them.
- Managing emotions: Remaining calm and avoiding panic so you can maintain sound judgment and make good decisions.
- Identifying assets: Noting what skills, abilities, and material possessions you have that will help you in this survival scenario.
- Perseverance: Knowing that you are capable of far more than you think you are and putting one foot in front of the other, proving it to yourself with every step.

The right mindset can make or break a situation. I have watched countless people succeed or fail when placed in emergency scenarios. Through the years I have tested what I consider the most essential strategies for surviving and thriving in extreme situations. I have documented and taught them to many people who have then used them successfully throughout their lives to cope with both physical (outdoor) hardship and challenges in their everyday lives. These strategies are very simple and have been talked about forever, yet many people fail to remember or use them when unexpected scenarios arise.

I believe these strategies can help ensure your success and survival too:

1. Educate yourself.
2. Be prepared.
3. Be positive.
4. Be adaptable.

It is wise to practice these strategies in your everyday life before you are tested in extreme conditions so they become second nature, not something you struggle to recall under stress. Let's take a closer look at each one.

EDUCATE YOURSELF

There is an old Latin saying: Scientia potential est—"Knowledge is power." But I prefer to think of it as "Knowledge is empowerment."

When I learn skills and am confident that I know what to do in a given situation, I feel empowered. The problem is **knowing** what is important to learn—that's where this book comes into play.

Learning new skills is always a good idea. They not only could prove useful in an unexpected situation but also will keep your mind and body active. Stretching yourself is the best way to improve yourself. You might read about something in the library or online. Take an evening course or join a class at the gym. Whatever it is, give it a go. Personally, I learn best when I have a teacher. I like someone to show me the ropes and then guide me as I try things out. There are many good teachers ready to impart their knowledge on a whole range of subjects. Look for them online, in the newspaper, or at your local community center.

A lot of people claim to be an expert at something, so be careful who you choose to be your mentor. I suggest going with a reputable company or someone who has good ratings and reviews rather than the cheapest price. I believe that when it comes to learning and being trained, you get what you pay for.

But don't just learn; practice the skills that you learn. Keep pushing yourself. Review your knowledge so you don't get rusty. That way when things go wrong, you are far more likely to remember what to do, having tried it and mastered it. It's also better to learn from mistakes when your life doesn't depend on it.

BE PREPARED

This next strategy is critical but something most people skip after educating themselves: Act on the information that you have just learned. You've done the research; now you need to do it in the real world. Book that CPR course, restock your first-aid kit, clean out your leaf-filled gutters, and learn how to navigate with a map and compass if your GPS batteries run out.

MAKE A LIST

Tasks can seem overwhelming at first, but a list turns a seemingly impossible project into manageable, bite-size chunks. Instead of listing a general task such as "fireproofing house," focus on clear, manageable actions:

1. Clean out gutters.
2. Cut away overhanging branches.
3. Clear away undergrowth around the house.
4. Make sure all hoses are functional and reach the roof.
5. Identify all roads that lead away from the house—prioritize an escape route.
6. Ensure all valuable possessions are in one place and easily transportable.

DO THE LIST

Individual items typically don't take much time to complete, meaning an entire list is often easily accomplished over a couple of weekends. Lists can also give us a sense of accomplishment as we tick things off, which encourages us to do more.

"Be prepared" isn't the Boy Scouts' motto for no reason. To be best prepared for the challenges and goals ahead of me, I try to do one thing on my list a day. It can be a full-day task or one that takes five minutes, depending on the time I have available, but simply setting and achieving that goal makes me the most prepared I can be for any given emergency.

BE POSITIVE

How often do we hear the saying "Be positive"? It comes out of the mouth of every motivational speaker I've ever heard, a catchphrase often thrown out without consideration for how to actually do it when times are tough. But it is popular for a reason: It really can turn your life and a situation around. Negative thinking has been shown to hasten giving up, especially in dire situations.

ACKNOWLEDGE THE SITUATION

The first thing I do when I find myself in a less-than-ideal situation is to acknowledge it. Let's say my neighborhood is flooded and I'm cut off from the shops. That's an awful situation to be in. But the first step is to face it head-on.

Denial of a situation is a short-term solution and often ends badly when reality sets in, so it is important to take a realistic look at what has happened and where you are. Unfortunately, many people get stuck dwelling on this point, but this is precisely when you need to move ahead.

Choosing to be negative about the situation you are in puts focus on, and enhances the perception of, the bad things about the circumstances. If you are in an emergency situation, you may be cold, tired, and hungry, but focusing on those things will not make them go away. Concentrating on them can make them seem worse than they are.

Choose instead to focus on what you do have, like the fact that you are alive and coherent. Looking at the things that are working for you will help you figure out how best to deal with the situation. It allows you to discover your strengths and then use them to your advantage.

ASSESS YOUR ASSETS

Take stock of what is working for you in the situation: Are you injured? Can you function? What resources do you have? The more things you can find that are in your favor, the less you will focus on the dire scenario.

There is a difference between focusing on the negatives and assessing your weaknesses. Focusing on the negatives is complaining that you are cold. Assessing your weaknesses is realizing that your electricity is out and the food in your freezer is thawing. Noting that this is a weakness in your survival plan will allow you to make a strategy and act. For example, start by using up perishable foods first to avoid waste, then transfer any remaining food from your fridge into your makeshift cooler (see page 199).

BE PRODUCTIVE

Something I ask myself when I am in extreme survival scenarios is, "What can I do today to make tomorrow better?" This not only helps me improve my current conditions but also keeps me proactive and in success mode. The problem-solving aspect of the question keeps my mind from wandering to dark places and distracts me from focusing on what I don't have.

BE ADAPTABLE

"The best laid plans of mice and men..." Unfortunately, even with the best of planning, sometimes things still don't turn out the way that we planned. Our well-rehearsed escape route may become blocked off or our water storage tanks may have leaked. Emergency scenarios are seldom ideal, which is why it's important to maintain flexibility in your thinking, even within a well-structured plan. This is where problem-solving skills come into play.

ASSESS THE SITUATION

After removing yourself from danger, the next thing to do is assess the situation. Consider your strengths and weaknesses within that scenario. Then figure out a way to make the strengths work for you.

TAKE THE TIME TO MAKE A PLAN

The next course of action is to make a plan based on your current predicament. By weighing the pros and cons of the situation, you will be able to determine the best course of action. By planning, you will avoid wasting time and energy trying avenues that will not be effective.

It may seem that taking the time to plan delays action and therefore prolongs an outcome. Yes, this may be the case sometimes (like when you need to move fast to avoid an oncoming obstacle), but typically sitting down and planning saves time in the long run. And it will make you feel proactive.

In an emergency situation, a feeling of helplessness can be just as deadly as the environment you are in. Your plan should be to assess the seriousness of the situation. Was it an earthquake that shook your house for a few minutes? Did your fight-or-flight mechanism take you to the street, where you joined your neighbors for a quick chat before returning to your home to straighten the pictures? Did a hurricane leave you without power and cut you off from the nearest town for three days? If that's the case, you will need to think about how to keep yourself warm, hydrated, and fed until power can be reestablished.

Your plan should take into consideration your personal "tool kit"—the lessons and skills you have learned through life. The more education and skills you bring into whatever situation life throws your way, the better your chances of coming out the other side with a positive result. Your personal tool kit could include:

- The knots that you know how to tie
- A first-aid course you took at work
- The experience you have purifying water when backcountry camping

You never know what may come in handy in any given situation, so count every skill as a bonus.

TAKE ACTION

From the second you realize things are looking grim, do what you can to improve your situation. This will not only increase your chances of success but also enable you to be in better condition when the crisis is over. So, once your plan is in place, take action.

PIVOT IF NECESSARY

It is important to remember that even though you have made a plan, it does not have to be set in stone. If it is not working for you, or if the situation changes, reassess and make a new plan. Don't stubbornly persist with a plan that is no longer working.

ATTITUDE CHECK

Your attitude in an emergency has a major impact on the outcome. While you can't always control the situation, you can control your mindset—and that makes all the difference. The key is recognizing that your attitude is changeable, and choosing to shift it in a positive direction is the first step toward success.

PART 1

SHELTER

WHEN THE POWER GRID GOES DOWN, your home becomes your life raft: a sanctuary where safety and comfort must be preserved against the uncertainties of the outside world. In this section, we explore how to maintain your home as a habitable refuge, ensuring it remains both a protective shelter and a comforting space for your family during challenging times.

In the absence of electricity and running water, simple measures like insulating your home can keep the cold at bay, while candles and lanterns provide illumination when darkness falls. A wood-burning stove or portable heater can offer both warmth and a means to cook hot meals, nourishing body and spirit alike. Beyond these essentials, maintaining a sense of normalcy becomes crucial. Crafting cozy spaces for your family to gather, share meals, and find solace can sustain morale. Simple comforts, like a favorite blanket or a cherished book, can provide much-needed reassurance and familiarity.

In preparing your home for these scenarios, we also consider the importance of defensibility, especially during natural disasters: reinforcing windows and doors, securing loose objects that can become hazards in high winds or earthquakes, and establishing a safe room where your family can retreat if needed.

By proactively addressing these aspects, your home becomes more than just a shelter; it becomes a symbol of resilience and comfort. With foresight and preparation, you can transform potential adversity into peace of mind, ensuring your family remains safe and secure, no matter what challenges arise.

PROJECTS

1 SET UP BACKUP POWER SOURCES

Having backup power sources is not just a luxury; it's a critical component of a comprehensive survival strategy. When the grid goes down—whether due to natural disaster, technical failure, or an unforeseen event—the absence of electricity can quickly escalate from an inconvenience to a crisis. Backup power sources such as generators, solar panels, or battery storage systems help to bridge this gap, ensuring that your home remains functional and your family stays safe and comfortable.

First and foremost, backup power provides continuity in essential services. It keeps vital appliances such as refrigerators and freezers running, preserving food supplies and medications that require refrigeration. In extreme temperatures, backup power can operate heating or cooling systems, protecting your family from the dangers of heatstroke or hypothermia.

Communication is another crucial aspect. During emergencies, staying informed and connected with the outside world can be life-saving. Backup power allows you to charge phones, radios, and other communication devices, facilitating your access to news updates, emergency alerts, and loved ones.

Moreover, backup power supports lighting and security systems, helping to prevent accidents in the dark and deterring potential intruders. It also enables the continued use of essential household tools, which may be necessary for repairs or adjustments during prolonged outages.

For people with medical needs, backup power is vital for operating life-sustaining equipment, such as oxygen concentrators or CPAP machines, without interruption.

Investing in backup power sources equips your home to withstand grid failures. By planning ahead and making sure these systems are ready to activate when needed, you can safeguard your family's well-being and enhance your home's self-sufficiency. In doing so, you transform uncertainty into preparedness, ensuring that your home remains a stable and secure haven in the face of adversity.

Several backup power sources can help ensure your home remains functional during an outage. We are going to look at the three most common: generators, solar power systems, and battery backup systems.

PORTABLE GENERATORS

A portable generator is a gas-powered machine that produces electricity and can be moved easily from place to place.

PROS: Affordable and easy to use; suitable for short-term power needs

CONS: Requires fuel, emits fumes, and can be noisy

PORTABLE SOLAR POWER SYSTEMS WITH BATTERY STORAGE

A portable solar power system with battery storage is a compact, mobile device that uses solar panels to capture sunlight and a battery to store the energy. It lets you power devices like phones, lights, or small appliances anywhere—without needing fuel or being plugged into the grid.

PROS: Renewable, quiet, and easy to set up

CONS: Dependent on sunlight and may have lower power output

BATTERY BACKUP SYSTEMS

A home battery backup system is a large rechargeable battery or series of batteries that store power for if the grid goes down. The system typically includes an inverter and charge controller to ensure a continuous energy flow.

PROS: Quiet, clean, and can be charged using grid power or renewable sources

CONS: Limited power duration and may require a larger setup for whole-home coverage

GUIDELINES FOR SAFE FUEL STORAGE

1. Use approved containers. Select containers rated for the specific fuel type (e.g., metal jerry cans or HDPE plastic fuel cans with proper seals).
2. Store in a cool, dry, ventilated area. Keep fuel away from heat sources, sparks, pilot lights, and direct sunlight.
3. Limit quantities. Store only what you realistically need; large volumes may require permits and fire-rated storage.
4. Label clearly. Mark each container with fuel type and the date stored.
5. Stabilize fuel. Add fuel stabilizers for long-term storage.
6. Rotate stock. Follow FIFO (first in, first out) practices to keep supply fresh.

TYPES OF FUEL

GASOLINE

- **Shelf life:** Stores for 3 to 6 months untreated, or up to 2 years with a stabilizer added.
- **Storage:** Keep in tightly sealed containers filled to 90 to 95% capacity to allow for expansion.
- **Risks:** Highly volatile and flammable; ethanol blends absorb water and degrade more quickly.
- **Tip:** Use non-ethanol gasoline when possible, as it lasts longer in storage.

DIESEL

- **Shelf life:** Stores for 6 to 12 months untreated, or up to 2+ years with stabilizers added.
- **Storage:** More stable than gasoline but susceptible to microbial growth ("diesel bug") in the presence of water.
- **Risks:** Water contamination and microbial growth.
- **Tip:** Store in opaque, tightly sealed containers and add a biocide for long-term storage.

KEROSENE

- **Shelf life:** Keeps for 5+ years when stored properly.
- **Storage:** Store in metal cans or approved plastic containers, away from moisture.
- **Risks:** Absorbs moisture over time and can degrade wick-based heaters or lamps.
- **Tip:** Use only water-clear, 1-K kerosene for long-term storage and heating applications; avoid mixing grades.

PROPANE/LPG

- **Shelf life:** Keeps indefinitely (does not degrade).
- **Storage:** Store in DOT-approved cylinders or tanks, always in an upright position.
- **Risks:** Leaks can accumulate in enclosed spaces; store outdoors or in well-ventilated sheds only.
- **Tip:** Check cylinder certification dates and inspect valves and seals regularly.

WHEN TO DO IT

Install backup power during the initial stages of creating your emergency preparedness plan. This ensures that your home is equipped to handle unexpected outages, providing peace of mind and security. An opportune time to integrate backup power systems is when you're already planning renovations or upgrades. This can simplify installation and potentially reduce costs.

PORTABLE GENERATORS

MATERIALS

- Portable generator with user's manual (see page 26)
- Fuel supply (see Tips)
- Fuel stabilizer (see Tips)
- Extension cords
- Generator cover or enclosure (see Tips)
- Engine oil (consult manufacturer's specifications; see Tips)
- Oil filter, air filter, and spark plug (for regular maintenance)
- Fire extinguisher
- Carbon monoxide detector

KEY ACTIONS

1. Select a suitable location. Place the generator outdoors in a well-ventilated area, at least 20 feet away from windows, doors, and ventilation intakes to prevent carbon monoxide poisoning. Ensure the location is dry and level, and use a protective cover if necessary to shield the generator from precipitation.
2. Fill the generator with the appropriate fuel (gasoline, diesel, or propane), as specified in the user's manual.
3. Check oil level and top up if necessary. Ensure the generator is on a stable surface to avoid spills.

4. Use heavy-duty, outdoor-rated extension cords to connect essential appliances directly to the generator. Alternatively, if your setup includes a transfer switch (see page 27), connect the generator to your home's electrical panel using the transfer switch.
5. Follow the manufacturer's instructions to start the generator. This usually involves turning on the fuel valve, setting the choke, and pulling the recoil cord or pressing the start button.
6. Once the generator is running, use extension cords to plug in the appliances or circuits you need to power, ensuring you do not exceed the generator's wattage capacity.
7. When you no longer need the generator, ensure all connected appliances are turned off and unplugged before shutting it down.

TIPS

- For a three-day emergency supply, you will need approximately 12 gallons of fuel (consult manufacturer's specifications).
- Fuel stabilizer is a chemical additive designed to slow the breakdown of gasoline or diesel during storage. It will extend the shelf life of your fuel from a few months to a year or more. Follow the manufacturer's instructions.
- Engine oil is a protective lubricant and may be required to coat moving parts, depending on manufacturer requirements.
- Ensure extension cords are long enough to reach from the generator's location to where power is needed and that they accommodate the number of prongs you require.
- Prioritize essential appliances like refrigerators, lights, and medical equipment.
- Regularly check fuel levels, oil, and overall operation.
- Follow the manufacturer's instructions to turn off the engine and close the fuel valve.
- After use, allow the generator to cool before refueling or storing.

CAUTIONS

- Always operate the generator outdoors in a well-ventilated area, at least 20 feet away from windows, doors, and ventilation intakes to prevent carbon monoxide poisoning.
- Keep the generator away from combustible materials and ensure it is placed on a stable, nonflammable surface.
- Follow the manufacturer's instructions for grounding the generator to prevent electrical shock.

HOW TO CHOOSE THE BEST GENERATOR FOR YOUR NEEDS

Not all generators are built the same. Whether you're dealing with a blackout, heading off-grid, or looking to keep your whole home running in an emergency, it's important to choose the right size for your needs—without buying too much or too little.

SMALL GENERATOR (1,000 to 2,000 watts/1 to 2 kVA)

- **Powers:** A fridge or freezer, a few lights, phone or laptop charging, a small TV
- **Good for:** Camping, short blackouts, keeping food cold

MEDIUM GENERATOR (3,000 to 5,000 watts/3 to 5 kVA)

- Powers: Fridge and freezer, lights, modem, a few power tools, possibly a small pump
- Good for: Running the essentials in a house during an outage

LARGE PORTABLE GENERATOR (6,000 to 9,000 watts/6 to 9 kVA)

- Powers: Fridge, freezer, lights, water pump, washing machine, small air conditioner
- Good for: Most-of-home backup (excluding large air conditioners or electric ovens)

STANDBY/FIXED GENERATOR (10,000+ watts/10+ kVA)

- Powers: Most of a house, including large air conditioners, electric stoves, and pumps.
- Good for: Full home or farm backup during long outages.

GENERATOR SIZING MADE SIMPLE:

- Camping or tiny backup → Small generator
- Household essentials → Medium generator
- Most of the house → Large generator
- Everything, automatically → Standby unit

- Protect the generator from rain and moisture to prevent electrocution and mechanical damage. Use a canopy or tent designed for generator use if necessary.
- Make sure all connected devices are turned off before starting the generator.
- Avoid overloading the generator and keep an ear out for unusual noises, which could indicate a problem.

WHAT IS A TRANSFER SWITCH?

A transfer switch is a small electrical panel that safely connects a generator to your home's wiring. Instead of running extension cords through doors or windows, a transfer switch lets you power selected household circuits—such as the fridge, lights, or water pump—directly.

WHY IT MATTERS

- **Safety:** Prevents dangerous "backfeeding" into the grid, which can injure utility workers or damage your generator.
- **Convenience:** With a single switch, your essential circuits run much like they normally would.
- **Code compliance:** In many areas, a transfer switch is legally required for whole-home generator hookups.
- **Control:** You choose which circuits stay on (e.g., kitchen or water pump) and which stay off (e.g., oven or clothes dryer).

TYPES

- **Manual transfer switch:** You start the generator and switch power sources yourself. This type is more affordable and sufficient for most emergency setups.
- **Automatic transfer switch (ATS):** Detects a power outage and switches to generator power automatically. This type costs more but provides seamless backup.

If you're planning to use a generator during outages, budget for a transfer switch in your setup. It's safer and turns a generator from a workaround into a reliable emergency power solution. Professional installation by a licensed electrician is strongly recommended.

- Store fuel in approved containers, away from living areas, and keep it tightly sealed. Refuel the generator only when it is turned off and cooled down to prevent fire hazards.
- Use heavy-duty, outdoor-rated extension cords with grounded plugs to connect appliances.
- Turn off and unplug all appliances before shutting down the generator. Follow the manufacturer's instructions for proper shutdown procedures.
- Perform regular maintenance to ensure safe operation, including changing the oil and cleaning air filters as recommended by the manufacturer, to keep the generator in good working condition.

SOLAR POWER SYSTEM WITH BATTERY STORAGE

QUICK RULES

- Think in 400 W panels.
- Battery is what gets you through the night; prioritize safe, deep-cycle, long-life batteries.
- Inverter must handle the biggest thing you turn on at once.

BARE ESSENTIALS (phones, Wi-Fi, a few LED lights)

- Panels: 1 × 400 W
- Battery (usable): 1 kWh
- Inverter: 300–600 W pure sine
- What you can do: Charge devices, run modem, 6–10 LED bulbs

FOOD SAFE (add a modern fridge)

- Panels: 2 × 400 W
- Battery (usable): 2 kWh
- Inverter: 1,000–1,500 W
- What you can do: Everything in "Bare Essentials" + run an efficient fridge 24/7

FOOD + COMMS + COMFORT (fridge, lights, Wi-Fi, laptop, TV)

- Panels: 3 × 400 W
- Battery (usable): 3 kWh
- Inverter: 2,000 W
- What you can do: Tier 2 + a laptop workday and a few hours of TV

SMALL HOUSEHOLD FLEX (fridge + lights + Wi-Fi + laptop/TV + short microwave/kettle use)

- Panels: 4 × 400 W
- Battery (usable): 4 kWh
- Inverter: 3,000 W
- What you can do: Tier 3 + brief high-watt bursts (1–2 boils or a quick reheat)

MATERIALS

- Solar panels with user's manual
- Battery storage (see Battery Backup Systems, page 32)
- Charge controller (see page 34)
- Inverter (see page 34)
- Cables and connectors
- Carrying case or protective cover

KEY ACTIONS

1. Choose an area to set up your solar panels. Make sure the location receives maximum sunlight throughout the day. (On average, you will need three to four hours of direct sunlight for solar panels to be effective. Solar panels will generate power in cloudy conditions, but their output is greatly diminished, so it's best to have a backup plan in winter months.) Ensure the area is free from obstructions like trees or buildings.
2. Unfold or set up the solar panels according to the manufacturer's instructions. Use any built-in stands or adjustable legs to position the panels at the optimal angle toward the sun.
3. Use the supplied cables to connect the solar panels to the charge controller. (This device regulates the current and voltage flowing to the battery, preventing overcharging.)
4. Use appropriate cables to connect the charge controller to the battery pack. Ensure all connections are secure and that you follow the polarity indicated on the devices.
5. If your system includes an inverter to convert DC to AC power, connect it to the battery. This is necessary if you plan to power AC appliances.

6. Allow the solar panels to charge the battery fully before use. The charge controller will prevent overcharging and indicate when the battery is full.
7. Use the built-in USB or DC ports on the battery pack to charge devices directly. For AC devices, plug them into the inverter's outlets (see page 28 for wattage limits).
8. Regularly check the charge controller and battery indicators to monitor energy production and storage levels. Some systems offer mobile apps for remote monitoring.

TIPS

- A charge controller regulates the voltage and current coming from the panels so your batteries get the right amount of charge without being overcharged or damaged. It also stops the batteries from sending power back into the panels at night.
- Find a portable system that includes USB or DC ports for direct device charging, eliminating the need for separate adapters.
- Look for panels and components that are weather-resistant for outdoor use.
- Opt for high-efficiency solar cells to maximize energy capture, especially in variable lighting conditions.
- Some systems allow for additional panels or batteries to be added, providing scalability for increased energy needs.
- Throughout the day, adjust the angle and orientation of the solar panels to track the sun for maximum efficiency.
- When panels are not in use, store them in a protective case. Keep the battery charged to maintain its health, and store it in a cool, dry place.

CAUTIONS

- Ensure panels and batteries are securely set up to prevent tipping or falling, especially in windy conditions. Use stands or mounts as recommended by the manufacturer.
- Use properly rated cables and connectors to prevent electrical hazards. Ensure all connections are secure and follow the correct polarity to avoid short circuits.
- Utilize a charge controller to prevent overcharging, which can damage the battery and reduce its lifespan. Regularly monitor the charge levels.
- Avoid exposing batteries and panels to extreme temperatures, as this can affect their performance and longevity. Depending on the battery type, they are far less flexible with extremes. Avoiding temperatures below 35°F and above 122°F is a good rule of thumb.
- Store batteries in a cool, dry place when not in use.
- Keep flammable materials away from batteries and solar equipment. Ensure proper ventilation to prevent gas buildup, especially for lead-acid batteries. Avoid keeping batteries in airtight spaces or areas without any airflow.

BATTERIES BY TYPE

Lithium-ion. Best for applications requiring long run times. Lightweight, high discharge rates, and minimal maintenance. Ideal for electric vehicles and portable electronics.

Lead-acid. Suitable for applications that require high capacity at a lower cost, such as solar energy storage and backup power systems. Run time may be limited due to discharge depth and weight.

Nickel-cadmium. While robust and capable of deep discharges, their self-discharge rate and memory can limit effectiveness in some applications, making them less common in new designs compared to lithium-ion.

BATTERY BACKUP SYSTEMS

MATERIALS

- Deep-cycle batteries such as lithium-ion, lead-acid, or nickel-cadmium, chosen based on capacity, efficiency, and lifespan (see page 31)
- Inverter/charger (see page 34)
- Battery management system (BMS) (see page 34)
- Charge controller (see page 34)
- Electrical wiring and connectors
- Monitoring system

KEY ACTIONS

1. Select a cool, dry, and accessible location to install the battery backup system. Ensure it is well-ventilated and away from direct sunlight or heat sources.
2. Securely mount the batteries on a rack or place them in a battery enclosure, ensuring they are stable and protected from physical damage.
3. Mount the inverter/charger close to the battery bank, following the manufacturer's instructions. Ensure proper ventilation around the inverter to prevent overheating.
4. Connect the batteries in series or parallel, depending on the system's voltage requirements. Ensure all connections are secure and follow the correct polarity.
5. If using solar panels, connect the charge controller to the battery system to regulate charging. Ensure all connections are secure and follow the manufacturer's guidelines.
6. Ensure the battery reaches full charge using grid power, solar panels, or another energy source before initial use. Monitor charging through the inverter or charge controller display.
7. Turn on the inverter and ensure it is set to automatically switch to battery power during a grid outage. Test this function to verify operation.

TIPS

- Regularly check the system's display or monitoring app to track battery charge levels, inverter status, and overall system health. Follow the manufacturer's maintenance schedule for battery health and longevity.
- Periodically inspect connections, clean terminals, and check for any signs of wear or damage.
- When not in use, system components should be stored in a safe, dry location to prolong their lifespan and ensure they are ready for future use.

CAUTIONS

- Follow proper charging protocols to avoid overcharging or deep discharging, which can shorten the lifespan of the battery.
- Ensure good ventilation for the battery backup system, especially if using lead-acid batteries that can emit harmful gases.
- Perform regular checks and maintenance on your battery system to ensure it's functioning properly.
- Consider having an alternative backup power source, such as a generator, in conjunction with your battery system for extended outages.

INVERTER

An inverter converts the electricity stored in batteries (DC = direct current) into the type of power most household appliances use (AC = alternating current).

WHY IT MATTERS:

- Batteries and solar panels produce DC power (similar to a car battery).
- Homes and household appliances run on AC power (like a wall outlet).
- Without an inverter, you're limited to DC devices or direct USB outputs from a solar panel, if available.
- With an inverter, you can run everyday appliances—fridge, lights, laptops, TVs, even a kettle—just like normal.

In short, an inverter makes stored battery or solar power practical for everyday household use.

CHARGE CONTROLLER

A charge controller regulates the voltage and current coming from the panels so your batteries get the right amount of charge without being overcharged or damaged. It also stops the batteries from sending power back into the panels at night.

WHY IT'S IMPORTANT:

- Prevents overcharging, which shortens battery life.
- Prevents over-discharging, which can kill a battery.
- Protects against voltage spikes from the panels.
- Improves efficiency by matching panel output to battery needs (especially with MPPT controllers).

BATTERY MANAGEMENT SYSTEM (BMS)

Many small, off-the-shelf lithium batteries (such as 12V 100Ah packs) include a built-in battery management system (BMS). Larger or DIY battery banks, however, require an external BMS to stay safe and reliable.

An external BMS is a separate unit that monitors and protects a group of batteries wired together, rather than being built into a single battery case. It's typically used in custom battery banks made from multiple lithium cells or modules.

WHAT IT DOES

- Monitors each cell or module in the bank.
- Controls charge/discharge by signaling the charger or inverter to stop if limits are reached.
- Provides balancing so all cells stay even in voltage.
- Adds safety cut-offs for over-voltage, under-voltage, over-current, or overheating.

Without a BMS, one weak cell could overcharge or over-discharge and destroy the whole bank (or worse, cause fire).

2 MAKE A DIY EMERGENCY LAMP (CANDLE OR OIL)

In times of unexpected blackouts, commercial lighting solutions may not be available, but materials for a homemade lamp are often found in most households. Additionally, creating your own lamp can be a cost-effective alternative to purchasing specialized emergency lighting products. With minimal expense and effort, you can have a functional light source at your disposal.

DIY emergency lamps can also be customized to fit your specific needs and preferences. You can choose the size and type of candle or oil to use, allowing you to tailor the lamp to your situation. This flexibility is especially useful for different environments, whether you're at home or in a camp setting.

Making your own candles can also be a fun activity, so I have added two projects for DIY candles, using wax or oil, so you will have an endless source of candles for your lamps.

WHEN TO DO IT

Make your DIY emergency candles or lamps prior to any emergency. Stock up before storm season, when there are far more power outages than any other time of year.

WAX CANDLE

MATERIALS

- Wax (paraffin, beeswax, or soy; see Tips)
- Candle fragrance oils (optional; see page 39)
- Colorants (optional)
- Wicks (see page 38)
- Heat-resistant containers or molds

TOOLS

- Newspaper or protective cover
- Heat-resistant gloves
- Double boiler or melting pot
- Stirring utensil
- Pencil or chopstick or skewer
- Food thermometer
- Scissors
- Lighter or matches

KEY ACTIONS

A

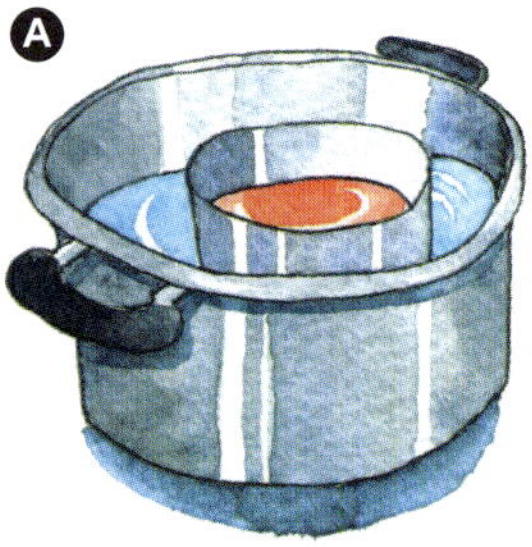

1. Prepare your workspace by laying down newspaper or another protective surface. Have a fire extinguisher nearby, and never leave melting wax unattended.
2. Place your wax in the top of a double boiler or heat-safe container placed inside a pot of boiling water. **(A)**
3. Stir the wax gently until it is fully melted.
4. If desired, add candle fragrance oil or dye to the melted wax (follow manufacturer's instructions). Stir well to combine.
5. Cut the wick to your desired length, ensuring it's a few inches taller than your container. **(B)**
6. Secure the wick to the bottom of the container using a small amount of melted wax. **(C)**
7. Lay the pencil horizontally across the top of your container. Then wrap or secure the wick around the pencil to hold it upright and roughly centered.
8. Carefully pour the melted wax into your container, leaving some space at the top. Ensure the wick remains centered as you pour. **(D)**

B

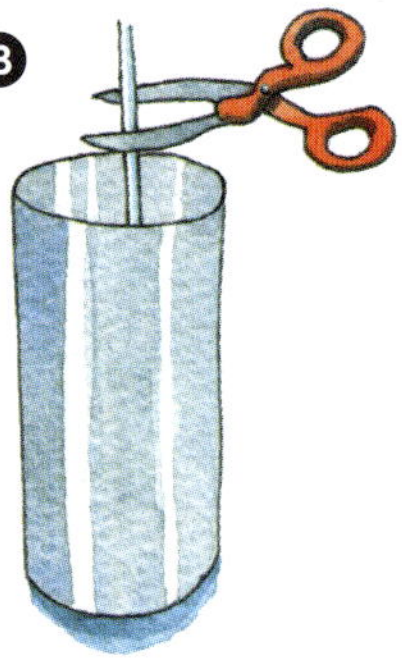

C

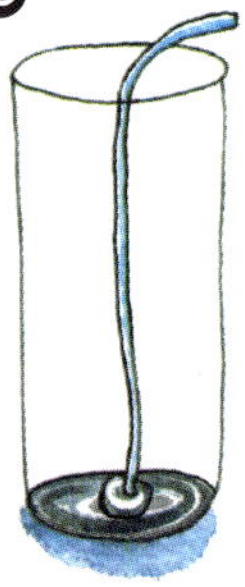

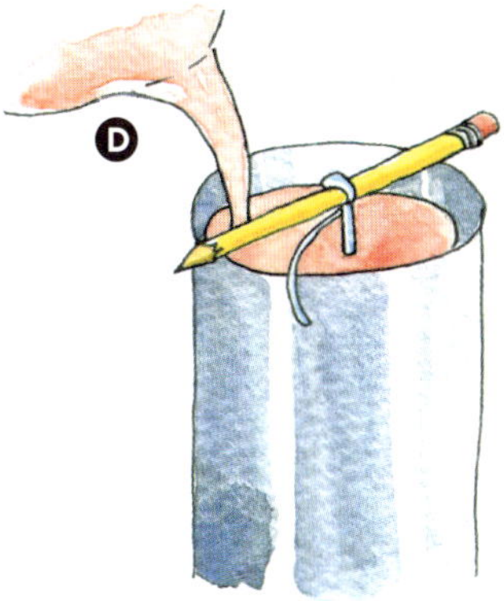

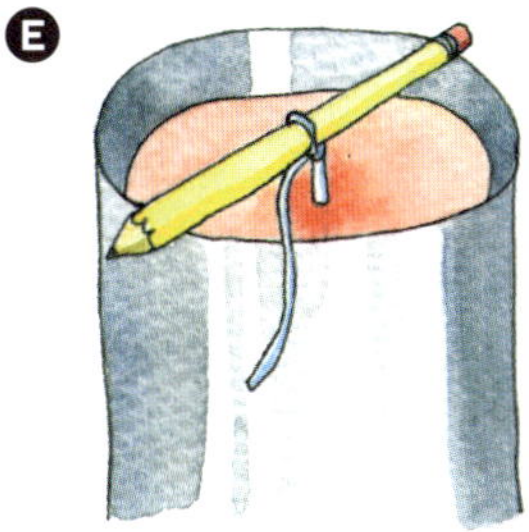

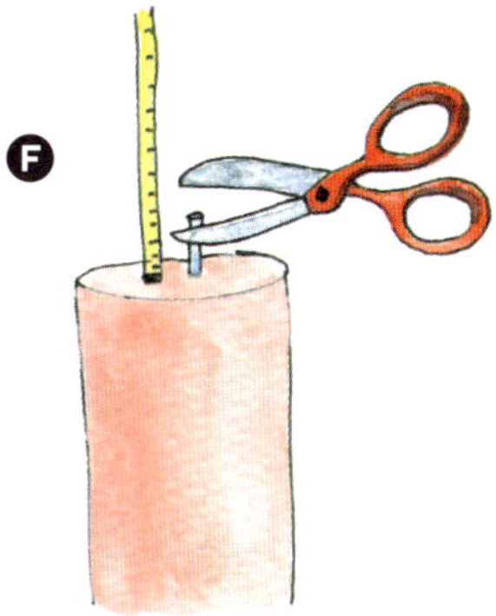

9. Leaving the pencil in place, let the candle cool and harden completely at room temperature. This may take several hours. **(E)**
10. Once the candle is fully cooled, trim the wick to about 1/4 inch above the surface. **(F)**
11. Let the candle cure for 1 to 2 days (or longer, depending on the wax type) to allow fragrance to bind.

TIPS

- For every 2 pounds of wax, you can make about four medium (8 oz.) candles.
- Always follow safety precautions when working with hot wax (see page 40).
- Experiment with different wax types, colors, and scents for unique candles (see page 38). When using scents, start by adding just a few drops, until you find a strength of scent that works just right for you.
- Using the wrong wick size can lead to excessive soot, smoke, or an unsafe burning candle (see page 38). Always test wicks to find the right match for your wax type.
- When working with hot wax, it helps to use a pouring pitcher for better control.

WAX CANDLE MAKING 101

Before you begin, review these basics—wick selection, wax temperatures, fragrance safety —to help you avoid common mistakes and get consistent results.

CHOOSING YOUR WICK TYPE

You have several wick options for candle making. The right choice depends on the type of candle you're making, the wax you're using, availability, and personal preference.

Wick Type	Pros	Cons
COTTON: Made from braided, plaited, or knitted cotton threads. The most common wick type for household candles.	• Reliable and widely available • Burns cleanly with minimal smoke when properly sized • Suitable for container candles, pillars, and tapers	• Can "mushroom" (form carbon buildup) if oversized • Requires regular trimming • May not be strong enough for very large candles without a core
WOOD: Flat or tube-shaped strips of treated wood that burn with a crackling sound.	• Creates a cozy, fireplace-like crackle • Wide flame produces a good melt pool for wider candles • Distinctive, modern appearance	• Can be difficult to light and keep lit • Requires careful sizing and testing for each wax type • Usually more expensive
ZINC-CORE: Cotton braid with a thin zinc wire core for added rigidity.	• Stays upright during pouring and burning • Well suited for container candles • Easy to center during pouring	• Can produce more soot and smoke • Less eco-friendly due to the metal core
PAPER-CORE: Cotton braid wrapped around a stiff paper filament.	• Stays rigid like zinc-core wicks • Produces a strong, steady flame • Good for large jar candles and pillars	• Can burn hotter and faster, shortening candle life • May produce more soot than all-cotton wicks
HEMP: Braided hemp fibers, sometimes coated in beeswax.	• Strong, slow-burning, and very natural • Well suited for beeswax candles • Often marketed as eco-friendly and chemical-free	• Not always easy to find • Can curl or lean if poorly made • Flame may be less stable than cotton

BEST WICK TYPES BY WAX

Wicks are typically cut to fit the container or mold, then trimmed to about ¼ inch before lighting.

Wax Type	Recommended Wick Types	Typical Wick Length
PARAFFIN WAX	Cotton (flat braid), zinc-core, paper-core	¼ inch
SOY WAX	Cotton (square braid), wood, hemp	¼ inch
BEESWAX	Cotton (square braid), hemp, thick flat braid	¼ inch

WAX MELTING AND POURING TEMPERATURES

CAUTION: If wax is overheated, it can ignite. This temperature is called the "flash point" and poses a serious safety risk.

Wax Type	Melting Point Range	Ideal Pouring Temperature
PARAFFIN WAX	115–154°F	150–160°F
SOY WAX	120–180°F	130–150°F
BEESWAX	144–149°F	160–170°F

CAUTION: Never heat wax anywhere near these temperatures. Use a double boiler or temperature-controlled wax melter, and always monitor with a thermometer. Most candle making occurs well below 200°F, keeping you safely within limits.

FRAGRANCES TO AVOID

Not all fragrance oils are safe for candles. Some can smoke, evaporate, or ignite due to low flash points.

- Essential oils with very low flash points (below 140°F) can be hazardous during melting and pouring. Examples include citrus oils (orange, lemon, lime, grapefruit), pine, eucalyptus, clove, thyme, and cinnamon leaf.
- Alcohol-based fragrances (such as perfumes or body sprays) are highly flammable and not candle-safe.
- Cooking extracts (such as vanilla or peppermint extract) are alcohol-based and unsuitable for candle making.

Use fragrance oils specifically labeled "candle safe." These are formulated to tolerate candle-making temperatures and bind properly with wax.

If using essential oils, choose those with higher flash points (above 170–180°F) and test in small batches. Lavender, cedarwood, patchouli, and candle-grade peppermint generally perform better.

CAUTIONS

- Melted wax can reach high temperatures and cause burns (see page 39). Always handle hot wax with care, using heat-resistant gloves and tools.
- When melting wax, ensure you're in a well-ventilated area. Some waxes, such as paraffin, can release fumes that may be harmful if inhaled.
- Keep other flammable materials away from your workspace. Never leave melting wax unattended, as it can ignite if overheated.
- Be cautious when adding fragrances or dyes. Some may be flammable or could cause allergic reactions (see page 39).

OIL CANDLE

MATERIALS

- Heat-resistant containers (see page 42)
- Oil (vegetable oil, olive oil, coconut oil, or lamp oil; see Tips)
- Candle fragrance oils (optional; see Tips)
- Cotton string wick
- Wick holder (optional; see Tips)
- Wick weight (optional; see page 42)
- Lighter or matches

KEY ACTIONS

1. Select a clean, heat-resistant container. Ensure it is dry and free from any residues.
2. Pour your chosen oil into the container, filling it about halfway so the wick will be sufficiently submerged and burn efficiently. **(A)**
3. Create the wick if not using pre-made: Cut a length of cotton string 4 to 6 inches long. Fray

one end to help it absorb the oil better (frayed end is top of wick, above oil, for lighting).

4. Place one end of the wick into the oil, ensuring that it reaches the bottom of the container. The other end should stick out ½ to 1 inch above the oil to serve as the flame source. **(B)** If the bottom of the wick floats, you can place a small weight at that end to keep it submerged (see page 42).

B

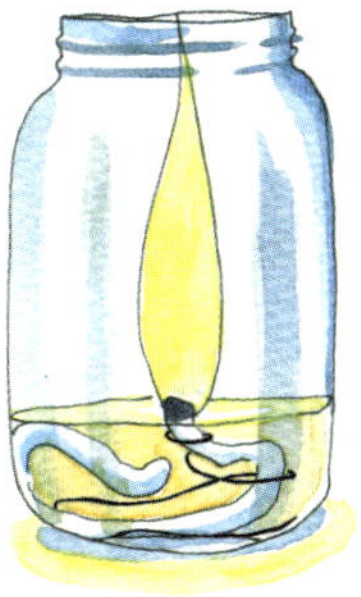

TIPS

- Once lit, the oil lamp will burn for several hours, depending on the size of the container and the amount of oil used. Never leave a candle unattended.
- Choose a high-quality oil like olive oil, coconut oil, or specialized candle oils for better burn quality and fragrance.
- The amount of oil required is equal to half the size of the container you choose. The larger the container, the more oil required and the longer the wick needed.
- Ensure the wick is centered and secured properly by the wick holder. This helps achieve an even burn and prevents wick leaning.
- If adding scent, use fragrance oils specifically designed for oil candles and follow the manufacturer's instructions. They are specifically formulated to burn cleanly and safely within the oil without clogging a wick or causing other issues. These oils are created to be combustible, producing a lasting scent, and are more stable than essential oils, which evaporate too quickly due to their volatility.
- Before making a large batch, make a small test candle to assess burn time, scent, and overall performance.
- Store your finished candles in a cool, dark place to preserve the oil's quality and extend shelf life.

OIL CANDLE MAKING 101

Before you begin, review these guidelines for choosing safe containers and keeping your wick properly anchored. A stable setup helps oil candles burn evenly and safely.

HEAT-RESISTANT CONTAINERS

Glass

- Tempered glass jars (such as Mason or canning jars) are designed to withstand heat and are less likely to crack.

Metal

- Stainless steel cups or tins are durable and shatterproof, but can become hot to the touch.
- Brass or copper vessels are traditional, attractive, and very durable.
- Aluminum containers are lightweight, inexpensive, and tolerate heat well.

Ceramic and Stoneware

- Ceramic bowls or mugs can be heat-resistant if fully glazed and nonporous.
- Stoneware or clay pots hold heat well but must be fully sealed or glazed so oil cannot seep through.

Other Safe Options

- Small enamelware mugs or dishes
- Recycled candle containers

Caution: Avoid thin glass (wine glasses, drinking tumblers), plastic, or untreated clay. These materials can crack, melt, or fail when heated.

WICK WEIGHTS

Oil candles require the wick to stay upright and centered in liquid fuel. Adding a little weight at the base keeps the wick anchored.

Common household options:

- Metal washers or nuts (steel or brass, with a center hole)
- Coins (stacked and glued to the wick base)
- Small fishing sinkers (compact and heavy, easy to attach)
- Glass or ceramic beads (heat-safe and won't corrode in oil)
- Flat stones/pebbles (smooth and thoroughly cleaned)

HOW TO ATTACH THE WICK

1. **Thread-through method:** Pass the wick through the center hole of a washer, nut, or bead, then knot it underneath.
2. **Glue or wax seal:** Use a dab of hot glue to fix the wick to the weight.
3. **Anchor tab:** Use a store-bought wick tab (small metal disc with a hole) and glue that to a coin or washer.

Tips

- Keep the weight small and flat so it sits securely on the bottom.
- Make sure the wick stays centered in the container for even burning.
- Pre-measure the wick length (½–1 inch above the oil surface) before securing.

WICK HOLDER

A wick holder prevents the top of the wick from sinking into the oil.

- **Lid method:** Using a hole punch or reamer, punch or drill a small hole in a jar lid and thread the wick through it.
- **Wire method:** Bend a piece of wire into a spiral to hold the wick upright, or shape it into a hook that rests on the jar's rim and supports the wick.

CAUTIONS

- Glass, metal, or ceramic containers work best. Avoid plastics or anything flammable.
- Use oil candles in well-ventilated areas to avoid the buildup of smoke or fumes.
- Avoid highly volatile or flammable liquids like gasoline or alcohol.
- Never leave burning candles unattended, and keep them away from flammable materials.
- To extinguish the flame, blow it out gently or use a snuffer to avoid splashing oil.

DIY EMERGENCY LANTERN

MATERIALS

- Glass jar, metal lantern, or any heat-resistant container
- Candle (oil or wax)

KEY ACTIONS

1. Select a clean, heat-resistant glass jar or container that is large enough to hold your candle. Ensure it is transparent enough to cast light.
2. Place the candle at the bottom of the jar or container.
3. Light the candle using matches or a lighter when you're ready to use it.

3 ESTABLISH A NON-ELECTRIC HEAT SOURCE

During freezing temperatures, a non-electric heat source, such as a woodstove, propane heater, or kerosene heater, can provide the necessary warmth to prevent hypothermia and related health risks in situations where power outages may occur due to natural disasters, extreme weather conditions, or other emergencies. These heat sources not only help keep individuals warm but also serve as practical tools for cooking. The ability to prepare food and boil water significantly reduces the risk of foodborne illnesses and ensures access to safe drinking water, both of which are essential for health and survival.

Relying solely on electric heating can leave individuals vulnerable during power outages, creating a sense of dependency on the electrical grid. Establishing a non-electric heat source helps eliminate this reliance, offering greater self-sufficiency and peace of mind during uncertain situations. When properly maintained and fueled, these alternative heating options, such as woodstoves, can operate for extended periods, providing consistent warmth even in prolonged emergencies.

Moreover, non-electric heating solutions are often versatile, suitable for both indoor and outdoor use. For instance, portable propane heaters can be easily relocated as needed, offering flexibility to adapt to changing conditions or environments.

There are several non-electric heat sources available, but I am going to focus on the following popular options that are relatively easy to set up and maintain.

Woodstove

Fireplace

Propane heater

Kerosene heater

WHEN TO DO IT

Include a non-electric heat source as part of your emergency preparedness plan before an emergency occurs, making sure you have the necessary equipment and fuel on hand. Regularly inspect, maintain, and test your heating options to ensure they remain functional and safe when you need them most.

INSTALLING A WOODSTOVE

MATERIALS

- Woodstove
- Chimney or flue pipe
- Wall heat shield (if needed)
- Noncombustible hearth pad (required under and in front of the stove)
- Smoke and carbon monoxide detectors

TOOLS

- Tape measure
- Carpenter's square
- Level
- Chalk line or pencil
- Drill/driver with bits
- Jigsaw or reciprocating saw
- Hole saw
- Tin snips
- Screwdrivers (flat and Phillips)
- Wrenches/spanners
- Hammer
- Masonry drill bits
- Work gloves
- Safety glasses
- Dust mask/respirator
- Ladder
- Caulking gun

KEY ACTIONS

1. Before installing a woodstove, check with your local building authority or fire marshal. Most areas require permits and inspections for woodstove installations. You'll need to comply with local building codes, fire codes, manufacturer clearances, and insurance requirements.
2. Select a suitable location for the woodstove, ensuring it is away from flammable materials and adheres to local building codes. Follow the manufacturer's specifications for clearance on all sides of the stove, including behind and above. Use heat shields, if necessary.
3. Install a heat-resistant hearth pad beneath the stove (follow local building codes). The pad should extend a specific distance around the stove to catch any embers.
4. Follow the manufacturer's instructions to assemble the chimney or flue pipe components. Ensure it is compatible with your woodstove.
5. Attach the chimney or flue pipe to the stove's outlet, securing it with clamps or screws, as necessary.
6. Ensure the chimney rises vertically and extends above the roofline (follow local building codes for height and clearance). Be sure to install a chimney cap and spark arrestor to prevent debris, animals, and precipitation from entering the chimney and to help reduce the risk of chimney fires.
7. Use high-temperature silicone sealant or chimney cement to seal any joints in the flue pipe to prevent leak.

8. Install smoke and carbon monoxide detectors or ensure existing detectors are in good working order.

9. Schedule a professional inspection (if required). In some jurisdictions, you'll need a certified inspector to sign off before the stove is used—especially if it affects your home insurance. Note that failing to notify your insurer about the installation of a woodstove could void your homeowner's policy or affect claims.

10. Before using, ensure all connections are secure and conduct a visual inspection for any signs of damage. Test the chimney draft by lighting a small piece of paper in the stove to check for proper airflow.

USING A WOODSTOVE

MATERIALS

- Seasoned firewood
- Kindling
- Newspaper or fire starter
- Long matches or lighter

TOOLS

- Fireplace tools (e.g., small shovel, poker stick, broom)
- Ash bucket
- Heat-resistant gloves (optional)

KEY ACTIONS

1. Use only dry, seasoned hardwood cut to a size that fits comfortably in your fireplace. ("Seasoned" wood has been dried naturally over time to reduce its moisture content to below 20 percent, making it suitable for burning in fireplaces and stoves. Aim for wood that has been dried for at least six months.) Avoid using green (i.e., freshly cut) or wet wood, as it produces more smoke and creosote buildup (see page 53).

2. Layer kindling and paper in your stove's firebox. Place crumpled newspaper or fire starters at the base, then stack small, dry kindling on top in a tepee shape. Leave enough space between pieces to allow airflow.
3. Before lighting anything, fully open the air-intake control to ensure proper airflow. A closed damper will trap smoke in your home.
4. Light the paper or fire starter using a long match or lighter. Once the kindling fully catches fire, gradually add larger logs, ensuring proper airflow.
5. Close the door once the fire is burning well.
6. Adjust airflow controls to manage the burn rate and heat output. Avoid closing the air too much too early—this can cause creosote buildup (see Tips).

TIPS

- Regularly clean the ashes from your woodstove's fire box.
- Learn how to adjust the air-intake controls to regulate the fire's intensity. Opening the air supply increases the burn rate, while closing it slows the fire down.

CAUTIONS

- Keep a fire extinguisher nearby.
- Never leave a fire unattended.
- Clean the stove and chimney regularly to prevent creosote buildup, which can cause chimney fires.
- Maintain required clearances from walls, ceilings, and furniture as specified by the manufacturer.

INSTALLING A FIREPLACE

MATERIALS

- Firebricks or fire-resistant blocks
- Cement or fire-rated refractory mortar
- Gravel or sand
- Metal chimney pipe
- Fireplace screen
- Smoke and carbon monoxide detectors

TOOLS

- Tape measure
- Carpenter's square
- Level
- Chalk line or pencil
- Drill/driver with bits
- Jigsaw or reciprocating saw
- Hole saw
- Tin snips
- Screwdrivers (flat and Phillips)
- Wrenches/spanners
- Hammer
- Masonry drill bits
- Work gloves
- Safety glasses
- Dust mask/respirator
- Ladder
- Caulking gun

KEY ACTIONS

1. Before building, contact your local building authority. Most municipalities require a permit for permanent fireplaces. Building without a permit can violate fire codes and void home insurance. Always follow the manufacturer's installation instructions and local building regulations.
2. Select a safe and suitable location for your fireplace, ensuring it is away from flammable materials and has sufficient clearance.
3. Ensure there is adequate ventilation and consider proximity to a wall for the chimney.
4. Decide on the shape and size of your fireplace. A simple rectangular design works well for beginners. Sketch a plan to visualize the dimensions.

5. Clear the area where the fireplace will be built. Level the ground and create a foundation using gravel or sand to provide drainage and stability.
6. Begin by laying the firebricks or fire-resistant blocks in your desired shape as a base. Use a level to ensure they are even. Leave space for the opening where the fire will be visible.
7. Continue stacking the bricks or blocks to create the walls of the fireplace. Firebricks should form the entire inside of the firebox. Use cement or mortar to secure the bricks together. Allow the mortar to cure per the manufacturer's instructions.
8. Extend the floor area in front of the fireplace with additional bricks or stones to create a hearth. This should be made of fire-resistant material to catch any falling embers.
9. If your design requires a chimney, install a metal chimney pipe that extends above the roofline or outside wall to vent smoke. Ensure it is securely attached and anchored, and properly sealed to prevent leaks.
10. Allow the fireplace to dry completely according to the mortar or cement instructions. This may take several days.
11. Install smoke and carbon monoxide detectors or ensure existing detectors are in good working order.
12. Schedule a professional inspection (if required). In some jurisdictions, you'll need a certified inspector to sign off before the fireplace is used—especially if it affects your home insurance. Note that failing to notify your insurer about the installation of a fireplace could void your homeowner's policy or affect claims.
13. After curing is complete, **do a small test fire**: Check that smoke exits through the chimney, not into the surrounding area. Look for signs of leakage or poor draft.

TIPS

- Fire needs oxygen. Make sure there's adequate ventilation to prevent smoke from backing up or carbon monoxide buildup. In tight spaces, install an air vent near the firebox.
- Clean out ash after every few burns, check the flue for blockages, and have the chimney swept at least once a year to help prevent creosote fires.

CAUTIONS

- For any installation inside a house or permanent dwelling, hire a licensed builder or certified installer. They'll ensure it meets building codes, insurance requirements, and safety standards.
- Keep fire-safety gear nearby. A metal bucket with a lid for ash, a fire extinguisher, and a carbon monoxide detector are essential. Always have a way to smother a fire quickly.
- If you're setting up a fireplace in a shed, off-grid cabin, or survival shelter, you can do the installation yourself provided you maintain proper clearances, use fire-resistant materials, and ensure a safe chimney path. Follow all manufacturer guidelines closely.

USING A FIREPLACE

MATERIALS

- Seasoned firewood, cut to the appropriate size for your fireplace
- Kindling
- Newspaper or fire starter
- Long matches or lighter

TOOLS

- Ax
- Fireplace tools
- Ash bucket
- Heat-resistant gloves

KEY ACTIONS

1. Use only dry, seasoned hardwood cut to a size that fits comfortably in your fireplace. ("Seasoned" wood has been dried naturally over time to reduce its moisture content to below 20 percent, making it suitable for burning in fireplaces and stoves. Aim for wood that has been dried for at least six months.) Avoid using green (i.e., freshly cut) or wet wood, as it produces more smoke and creosote buildup (see page 53).
2. Layer kindling and paper in your fireplace. Place crumpled newspaper or fire starters at the base, then stack small, dry kindling on top in a tepee shape. Leave enough space between pieces to allow airflow.
3. Before lighting anything, fully open the air-intake control to ensure proper airflow. A closed damper will trap smoke in your home.
4. Light the paper or fire starter using a long match or lighter. Once the kindling fully catches fire, gradually add larger logs, ensuring proper airflow.

TIPS

- Regularly clean the ashes from your fireplace.
- Learn to adjust the air-intake controls to regulate the fire's intensity. Opening the air supply increases the burn rate while closing it slows the fire down.

CAUTIONS

- Keep a fire extinguisher nearby.
- Never leave a fire unattended.
- Clean the fireplace and chimney regularly to prevent creosote buildup, which can cause chimney fires.
- Proper ventilation is crucial to ensure smoke is safely directed outside.

HOW TO PREVENT CREOSOTE BUILDUP

Creosote is a flammable, tar-like substance that forms when wood smoke cools and condenses inside a chimney or flue. It's made up of unburned wood particles, gases, and moisture, and it accumulates over time, especially when burning green (wet) wood or operating a stove or fireplace at low temperatures. You can help prevent creosote buildup using the following ways:

- Burn only seasoned, dry wood.
- Ensure good airflow and hot-burning fires.
- Don't overload the firebox or dampen the fire too much.
- Have your chimney inspected and cleaned annually by a certified chimney sweep. (The National Fire Protection Association recommends at least once per year.)

PROPANE HEATERS

MATERIALS

- Propane heater unit (wall-mounted, portable, or freestanding), with instruction manual
- Propane tank
- Regulator
- Flexible propane hoses
- Fittings and connectors
- Mounting brackets (for wall-mounted or ceiling-mounted heater)
- Carbon monoxide detector
- Fire extinguisher appropriate for propane fires (typically rated for Class B fires)

TOOLS

- Wrenches
- Screwdrivers
- Drill

KEY ACTIONS

1. Select a suitable location for the heater, away from flammable materials and with access to fresh air for combustion. Check for proper clearance per the manufacturer's guidelines.
2. If your heater is wall-mounted, use a level to mark the position of the mounting bracket on the wall. Drill holes as needed and secure the bracket with screws.
3. Inspect the propane tank and connections for any signs of damage. Make sure the tank is full and properly sealed.
4. Connect the propane regulator to the propane tank's outlet. Ensure it is tightly secured but do not overtighten.
5. Attach one end of the flexible propane hose to the regulator and the other end to the heater's gas inlet. Use the appropriate fittings to ensure a secure connection.

6. If your heater requires venting, install the vent pipe according to the manufacturer's instructions. Ensure it leads outside to prevent carbon monoxide buildup.
7. Position the heater in the desired location. Secure it to the mounting bracket (if applicable). Follow the manufacturer's instructions for securing and leveling the unit.
8. If your heater has a thermostat, install it according to the manufacturer's guidelines. This may involve wiring or mounting it on the wall.
9. Before turning on the heater, check all connections for tightness. Use a mixture of soap and water to check for leaks; bubbles will form at any leaking points. Tighten as necessary.

10. Open the valve on the propane tank slowly. This will allow gas to flow to the heater.
11. Follow the manufacturer's instructions for igniting the heater. This may involve pressing a button or using a lighter for manual ignition.
12. If your heater has a thermostat, set it to your desired temperature.

TIPS

- Choose a heater that is appropriately sized for the space you want to heat. An undersized unit may struggle to heat the area, while an oversized one may lead to inefficiency.
- Follow local building codes and regulations regarding indoor propane heater installations.
- Use only hoses and fittings that are rated for propane.

- Use Teflon tape or pipe dope rated for gas connections to ensure a tight seal on threaded fittings. This helps prevent gas leaks.
- Regularly check and maintain your propane heater according to the manufacturer's recommendations.
- Plan for regular maintenance checks to ensure that the heater operates efficiently and safely. This includes cleaning and inspecting the unit and connections.

CAUTIONS

- Never leave the heater unattended while running.
- Use only heaters certified for indoor use. Outdoor-only heaters should never be used inside.
- Ensure proper ventilation is maintained in the area.
- Place a carbon monoxide detector in the same room as the heater to alert you to any dangerous gas levels.
- Have a fire extinguisher nearby that is appropriate for propane fires.
- If you are unsure about any aspect of installation, do not hesitate to consult a licensed professional. Safety should always be the top priority.
- Maintain manufacturer-recommended clearances between the heater and any combustible materials (curtains, furniture, paper, etc.).
- Never block air intakes or exhausts.

KEROSENE HEATERS

MATERIALS

- Kerosene heater unit, with user's manual
- Kerosene fuel
- Carbon monoxide detector

TOOLS

- Adjustable wrench
- Screwdrivers (flat and Phillips)
- Pliers
- Socket set or spanners (depending on bolt type)
- Utility knife (for packaging and hose trimming, if required)
- Work gloves
- Safety glasses
- Funnel

KEY ACTIONS

1. Select a suitable location for the heater that is at least 3 feet away from flammable materials (furniture, curtains, etc.); on a flat, stable surface to prevent tipping; and well-ventilated to allow for proper air circulation.
2. Follow the manufacturer's instructions to assemble the heater components. This may include attaching the fuel tank or any other parts.
3. Ensure the tank is cool. Using a funnel, pour clean, high-quality kerosene into the fuel tank. Avoid overfilling; leave some space to prevent spills.
4. Ensure the fuel tank is properly secured and that any caps or lids are tightly closed.

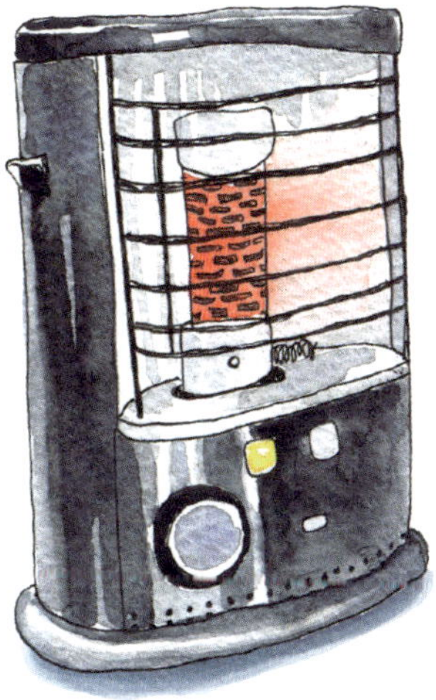

5. If your kerosene heater requires electricity for ignition or operation, plug the heater into a suitable outlet. Use an extension cord rated for the heater's power requirements, if necessary.
6. Place a carbon monoxide detector in the same room as the heater to alert you to any dangerous gas levels.
7. Follow the manufacturer's instructions for igniting the heater. This may involve turning on the fuel supply and using an ignition button or lighter.
8. Adjust the temperature settings according to your preferences.

TIPS

- Choose a kerosene heater that suits your space requirements, whether it's portable or stationary.
- Ensure the area has adequate ventilation (e.g., a cracked window or door slightly open) to prevent the buildup of fumes from the kerosene.
- Purchase clean, high-quality kerosene fuel, as using the wrong type can affect performance and safety.
- If the heater has a drip tray or pan to catch spills or leaks, ensure you have it in place.
- After starting the heater, monitor it for any unusual sounds, odors, or smoke.

CAUTIONS

- Never leave the heater unattended.
- Make sure you have a carbon monoxide detector in the room where the heater is installed to monitor for dangerous gas levels.
- Store kerosene in a properly labeled, sealed container, away from heat sources and children.
- Ensure the heater is turned off and cool before filling the fuel tank.
- Plan for regular maintenance checks to ensure that the heater operates efficiently and safely. This includes cleaning the heater and checking connections.
- Always refer to the manufacturer's manual for specific installation requirements and safety guidelines.
- If you have any doubts or concerns during the installation process, consult a professional.

4 ESTABLISH NON-ELECTRIC COOLING OPTIONS

The opposite of extreme cold conditions is extreme heat. Without electricity, cooling down can actually be more difficult than staying warm.

High temperatures can pose serious health risks, particularly for vulnerable populations such as the elderly, children, and those with preexisting health conditions. Non-electric cooling options help people regulate their body temperature, reducing the risk of heat exhaustion and heatstroke. They can allow individuals to stay comfortable and safe without relying on electricity, which can be vital in maintaining well-being during extreme heat.

Non-electric cooling methods are often more resource-efficient. They typically rely on natural elements, such as air flow and evaporation. Simple techniques—such as using battery- or propane-powered fans, creating cross-ventilation, or utilizing shaded areas—can effectively lower indoor temperatures without requiring a lot of resources, preparation, and, most importantly, electricity.

Simplicity and versatility are key attributes of non-electric cooling methods. Many of these techniques are easy to implement and can be adapted to different environments. For example, using wet cloths, taking cool baths, or employing reflective materials can all be effective strategies that cater to diverse situations.

Maintaining a comfortable temperature during emergencies can have a positive impact on mental well-being. Non-electric cooling methods contribute to reducing anxiety and discomfort, helping to restore a sense of normalcy and safety during challenging times.

Here are some of my favorite options. They may seem simple, but they can make a huge difference:

- During the hottest part of the day, keep windows and doors closed to prevent warm air from entering your home. Open them in the early morning or late evening when temperatures are cooler.
- Make use of battery-operated or solar-powered fans.

- Open windows on opposite sides of your home to allow for cross-ventilation. This helps create a natural airflow that can cool down indoor spaces.
- Hang damp curtains or sheets in front of windows. As the air passes through the wet fabric, it cools down before entering your living space.
- A quick splash of cold water can lower your body temperature significantly. If possible, fill a basin with cool water and soak your feet or hands.
- Place bowls of ice or cold water in front of fans to create a makeshift air conditioner. The fan will blow air over the ice, cooling it before it reaches you.
- Dress in light, breathable fabrics such as cotton to help your body stay cool. Avoid dark colors that absorb heat.
- Avoid cooking or using appliances that generate heat, such as ovens or stoves. Opt for cold meals or snacks, such as salads or fruits, to keep your kitchen cool.
- Use blankets, towels, or reflective materials to block sunlight from entering your home. You can also set up outdoor canopies or umbrellas to create shaded areas.
- Drink plenty of water to help regulate your body temperature. Staying hydrated is essential for maintaining comfort in hot conditions.
- Move to the floor of your bedroom to sleep if possible. The lower you can get, the cooler it will be.
- Place a thermometer both inside and outside the house so you can monitor temperatures.
- Have an emergency radio available to monitor advisories and predicted temperatures.

WHEN TO DO IT

It's crucial to prepare for potential power outages and cooling needs well in advance. Being proactive by creating a plan and gathering necessary supplies ensures you are ready when extreme heat or emergencies occur.

MATERIALS

- Battery-operated or solar-powered fans
- Coolers or ice packs
- Damp towels or cloths
- Reflective or blackout curtains
- Bottled water
- Portable battery chargers
- Lightweight bedding or cooling sheets
- Easy-to-eat items that don't require cooking
- Thermometer (indoors and outdoors)
- Emergency radio

KEY ACTIONS

1. Decide on methods you will use to stay cool, such as using battery-operated fans, taking cool baths, or creating shade. Identify rooms in your home with the fewest windows or doors, or perhaps the ones with windows facing away from the predominant seasonal direction of the sun. Downstairs rooms tend to be cooler than upstairs, as heat rises. Use your best judgment to avoid rooms that heat up quickly for any reason.
2. Collect supplies. Assemble and organize battery-operated or solar-powered fans, coolers, ice packs, damp towels, reflective or blackout curtains, bottled water, portable chargers, lightweight bedding, nonperishable food, an emergency radio, and a thermometer.
3. Hang reflective or blackout curtains in windows to reduce heat gain during the day.
4. Check for any gaps around windows and doors, and seal them to prevent hot air from entering.

5. Set up battery-operated or solar-powered fans in key areas of your home, such as living rooms and bedrooms, positioned to create cross-ventilation.
6. Freeze water bottles or use ice packs in coolers to keep food items cold and for personal cooling.
7. Fill bottles with water and keep them readily accessible to stay hydrated.
8. Stock up on nonperishable food items that don't require cooking, such as canned goods, nuts, fruits, and snacks. Organize these in an easy-to-reach location. Avoiding the use of the stove or oven will help keep the temperatures down.

TIPS

- Don't wait until you are too hot to try to cool down. It is easier to maintain a survivable temperature than to recover from overheating.

CAUTIONS

- High temperatures can lead to illnesses such as heat exhaustion and heatstroke. Vulnerable populations, including the elderly, young children, and those with preexisting health conditions, are particularly at risk.
- Without proper cooling, the risk of dehydration increases. It's essential to drink plenty of water to stay hydrated.
- Prolonged exposure to high temperatures can affect mood and mental health, leading to increased stress and anxiety levels. It's important to find ways to stay cool and calm.
- High heat can damage sensitive electronic equipment or supplies. Ensure that items are kept in cooler areas to prevent damage.
- If temperatures become unbearable and you cannot cool your home effectively, have a plan in place for evacuation to a cooler location.

5 SET UP AN OFF-GRID COOKING METHOD

The need for an off-grid cooking method becomes particularly pronounced during situations where the grid fails or when utilities become unreliable. In such scenarios, having an alternative cooking solution is crucial.

When the electricity grid goes down, access to conventional cooking appliances such as electric stoves, ovens, and microwaves is lost. An off-grid cooking method ensures that you can still prepare meals, providing sustenance and comfort when it is needed most.

In the absence of grid power, food preservation also becomes a pressing concern. Refrigeration relies on electricity, and without it, perishable items can spoil quickly. Off-grid cooking methods allow you to use nonperishable food supplies effectively and prepare meals without the need for refrigeration, helping to reduce food waste during outages.

It also allows you to maintain a sense of normalcy and routine, even in challenging circumstances. The ability to cook meals can boost morale and provide a vital psychological sense of control during stressful times.

WHEN TO DO IT

Incorporate off-grid cooking methods into your preparedness plan as a proactive measure before any potential disruptions to the grid occur. Ideally, establish these systems during your initial planning, allowing time for practice and familiarity with alternative cooking techniques ahead of any crisis.

There are many effective off-grid cooking methods for preparing meals without relying on traditional electricity or gas sources. Some popular examples covered in this book are the solar oven, woodstoves, and rocket stoves. Some others that I recommend are:

- Campfires
- Charcoal grills
- Propane stoves
- Alcohol stoves
- Earth or pit oven

There are many varied methods of setting up a **fire lay** (the structure made with kindling). Here I'll provide two different examples: the tepee lay and the log cabin lay. When it comes to making a fire pit, some people like to dig a hole to contain the fire, while others make a ring of stones. I personally like to have a clearly defined fire area so that the fire doesn't sprawl, but that's a matter of personal choice. Just remember: If you line your fire with rocks, you need to make sure that they are not river rocks. Wet rocks or river rocks tend to explode when heated and can cause nasty injuries. To avoid this, bang the rocks together and discard any that sound hollow or are brittle.

An important note on firewood: How well it burns will depend on where you are and the types of wood you have access to. As a general rule, hardwoods (such as oak and beech) burn hotter and longer than softwoods (such as aspen and cottonwood) and produce bigger coals.

THE TEPEE LAY

MATERIALS

- Tinder: very small, dry, flammable material (e.g., dry grass, bark shavings, dryer lint)
- Toothpick-sized twigs: very thin kindling, easy to ignite
- Pencil-sized twigs: slightly thicker kindling to build heat
- Thumb-sized sticks/branches: small fuelwood to sustain the fire
- Larger logs for later, once fire is well established (optional)
- Fire starter (matches, lighter, ferro rod)
- Water or fire extinguisher

KEY ACTIONS

1. Choose and prepare your fire site: Find a safe, clear area free of flammable debris, overhanging branches, and grass. Use a fire ring if one is available. If not, create a fire pit by clearing a 3-foot diameter circle and surrounding it with rocks.
2. Gather your fire materials: You'll need tinder, kindling, and firewood.

3. Build the tepee structure:

- Place a bundle of tinder in the center of your fire pit.
- Arrange toothpick-sized twigs over the tinder, leaning them together at an angle to form a small tepee structure.
- Around the outside of this, carefully add pencil-sized twigs, leaning them against the first layer to build a larger tepee.
- Add a few thumb-sized branches around the structure, also leaning inward like a tent frame. Leave a small opening or "door" on the side where you'll light the fire.

4. Light the fire: Place your heat source (match or lighter) under the tepee, touching the tinder. Light the tinder and allow the flames to rise naturally. As the fire burns upward, the tinder lights the toothpick twigs, the toothpicks ignite the pencil-sized twigs, and the pencil twigs build enough heat to ignite the thumb-sized branches.

5. Maintain the fire: Add more fuelwood gradually, starting with sticks about the size of your wrist. Don't overload the fire—build it up slowly. Always keep an eye on the fire, and never leave it unattended.

6. When finished, extinguish the fire properly: Let the fire burn down to ash, if possible. Douse with water, stirring the ashes with a stick to expose embers. Add more water until no hissing sounds remain and the ashes are cool to the touch. Double-check the area before leaving.

THE LOG CABIN LAY

MATERIALS

- Tinder: dry, flammable material (e.g., dry grass, birch bark, or cotton balls)
- Toothpick-sized twigs: very fine kindling
- Pencil-sized twigs: slightly thicker kindling
- Thumb-sized branches: small fuelwood
- Two larger logs: these form the base of the "cabin"
- Fire starter (matches, lighter, ferro rod)
- Water or fire extinguisher

KEY ACTIONS

1. Choose and prepare your fire site: Find a safe, clear area free of flammable debris, overhanging branches, and grass. Use a fire ring if one is available. If not, create a fire pit by clearing a 3-foot diameter circle and surrounding it with rocks.

2. Gather your fire materials: You'll need tinder, kindling, and firewood.

3. Lay the foundation: Place two large logs on the ground, parallel to each other, 12 to 16 inches apart. These are the first "walls" of the cabin. Between the logs, scrape out a shallow hole and place your tinder bundle in it.

4. Build the log cabin structure:

    - Lay a small handful of toothpick-sized twigs across the hole and tinder—half going in one direction, the other half going at right angles (criss-crossed like a grid).

    - On top of the large base logs, place the pencil-sized

twigs, again in a crisscross pattern (alternating directions with each layer).

- Continue with the thumb-sized branches in the same crisscross pattern, forming the "walls" of your cabin as it gets taller. You can place larger fuelwood on top early—the stable frame supports the weight and won't collapse like a tepee can.

5. Light the fire: Use a match or lighter to ignite the tinder in the center. The fire will spread upward, catching the toothpick twigs first. As each layer ignites, it creates enough heat to light the thicker layers above. The open structure allows good airflow, and the crisscrossed wood provides a steady heat buildup.

6. Maintain the fire: Once the fire is established, add more thumb- to wrist-sized logs on top or around the fire as needed. Keep the fire at a controlled size, and never leave it unattended.

7. When finished, extinguish the fire properly: Let it burn down to ash, if possible. Douse with water, stirring the ashes with a stick to expose embers. Add more water until no hissing sounds remain and the ashes are cool to the touch. Double-check the area before leaving.

TIPS

- Make sure to have a bucket of water and a shovel nearby for safety.
- Try to find a dry location out of the wind.
- If the ground is damp, lay down a branch platform to keep your fire base dry. Make the platform about a foot square in size, placing the branches directly on the wet ground to create a raised, dry base for the fire.
- Don't pack the wood on too thickly or too thinly; too much wood will smother the flame and too little won't provide enough fuel and the flame will burn out.

- If the fire struggles, gently blow at the base to add oxygen. Feed in more kindling if needed.
- It's always good to practice cooking on a fire before you need to. Plan fun BBQ lunches or interesting snacks to make with the family to get started.
- Make a permanent fire pit if you plan on having this as a backup cooking method for emergencies.

CAUTIONS

- Always check local regulations regarding campfires, as some areas may have restrictions or fire bans.
- Clear the area around the fire pit of flammable materials to prevent accidental fires.
- Make sure that you have fire-retardant gloves as well as fire-specific cooking utensils such as long tongs and prongs to protect you from burns while cooking.

CHARCOAL GRILL

MATERIALS

- Grill (portable grill or a larger stationary unit; typically includes a cooking grate, a charcoal grate, and ventilation controls)
- Grill brush
- Crumpled newspaper or lighter cubes
- Charcoal
- Fire starter (matches, lighter, ferro rod, etc.)
- Heat-resistant gloves
- Fire extinguisher or water

KEY ACTIONS

1. Ensure the grill is clean and free of ash from previous use. Use a grill brush to clean the grates, if necessary.
2. Place crumpled newspaper or fire-starter cubes in the bottom section of the grill.
3. Arrange the charcoal.
4. Light the newspaper or fire starter through the holes at the bottom.
5. Allow the charcoal to ignite for ten to fifteen minutes, until it's covered with white ash.
6. Once the charcoal is ready, place the cooking grate back onto the grill. Allow it to heat up for a few minutes.
7. When finished grilling, allow the charcoal to burn out completely. To extinguish, close the vents to cut off airflow, which will eventually put out the fire. Alternatively, if you want to douse it, carefully pour water over the coals and stir to ensure they are fully extinguished.

TIPS

- Close the grill lid to retain heat and smoke.
- After the grill has cooled down, clean the grates with a grill brush and dispose of the ash properly (in your garden, compost, or green bin).
- For added flavor, soak food-grade wood chips in water and add them to the charcoal for smoking.

CAUTIONS

- Always have a fire extinguisher or a bucket of water/sand nearby in case of flare-ups.
- If using lighter fluid, do not add it to already-lit coals, as it can cause dangerous flare-ups. Always follow the manufacturer's instructions.
- Use long-handled tools to avoid burns and splatters. Ensure all tools are heat-resistant and in good condition.
- Never leave a lit grill unattended, as this can lead to uncontrolled flames or accidents.

PROPANE STOVE

MATERIALS

- Propane stove (portable single or double-burner stove, depending on your cooking needs; you can also find small, camping-style burners that take up minimal space and can fit in your go bag; see page 227.)
- Propane fuel
- Propane canisters
- Larger propane tanks
- Regulator and hose
- Ignition source

KEY ACTIONS

1. Select a well-ventilated outdoor area away from flammable materials, structures, and overhanging branches.
2. Place the propane stove on a stable, flat surface, ensuring it's secure and won't tip over.
3. Ensure that the burner knobs are in the "Off" position before turning on the propane.

4. If using a propane canister, ensure it is tightly and correctly connected to the stove without cross-threading. If using a larger tank, connect the regulator and hose securely before opening the valve.

5. Turn on the gas valve slowly on the propane tank or canister. Do not lean over the stove while lighting it. Keep your face and body at a safe distance.

   - If the stove has a built-in igniter, press the ignition button while turning the burner knob to the "On" position. You should hear a click and see a flame.
   - If there's no igniter, use a long lighter or matches to ignite the burner while turning the knob to the "On" position. Stand back and ensure your hands are clear of the flame.

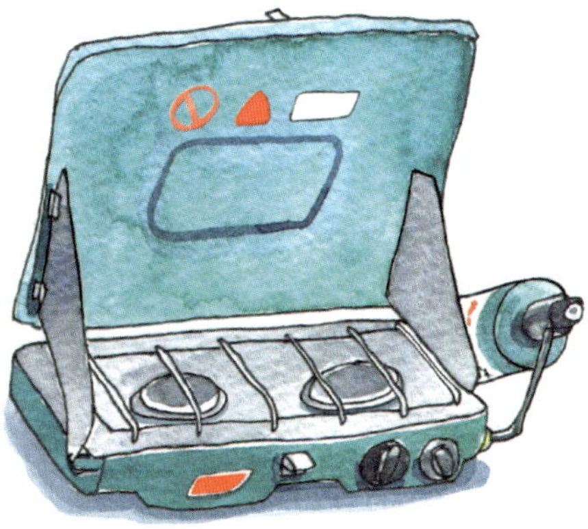

6. Once the burner is lit, adjust the flame to your desired heat level by turning the burner knob. A higher flame is suitable for boiling, while a lower flame is better for simmering.

7. Monitor the stove during use. Never leave a lit stove unattended.

8. Once you're finished cooking, turn off the burner knob to extinguish the flame. If using a tank, close the gas valve on the propane tank as well.

TIPS

- If you are using the stove outdoors, a windscreen can help protect the flame from wind, allowing for more efficient cooking.
- Allow the stove to cool down completely before cleaning. Use a grill brush to clean the grates and a damp cloth to wipe down surfaces.

CAUTIONS

- Safely store the propane canister or tank in an upright position in a well-ventilated area, away from heat sources. Ensure all connections are secure before storing the stove.
- Avoid using a propane stove indoors or in enclosed spaces due to the risk of carbon monoxide buildup.

ALCOHOL STOVE

An alcohol stove is a small, portable stove that uses denatured alcohol or other liquid alcohol fuels (like ethanol or methanol) to produce a flame for cooking or boiling water, particularly in outdoor, emergency, or minimalist situations. They come in various designs, including simple screw-top models and more complex ones with multi-burner options. Choose the one that best suits your needs (ask yourself if you need a simple temporary stove or a multi-burner stove that's less portable).

Fuel options include denatured alcohol (most commonly used with alcohol stoves and readily available at hardware stores or camping supply stores), isopropyl alcohol (commonly known as rubbing alcohol; ensure it is at least 70 percent concentration for effective burning), and methanol or methylated spirits. Some work more effectively than others. Use the fuel specifically recommended by the manufacturer, but know that any of the alcohols mentioned should work in a pinch.

MATERIALS

- Alcohol stove
- Fuel
- Fuel container (see Tips)
- Ignition source (matches or lighter)
- Water or fire extinguisher

KEY ACTIONS

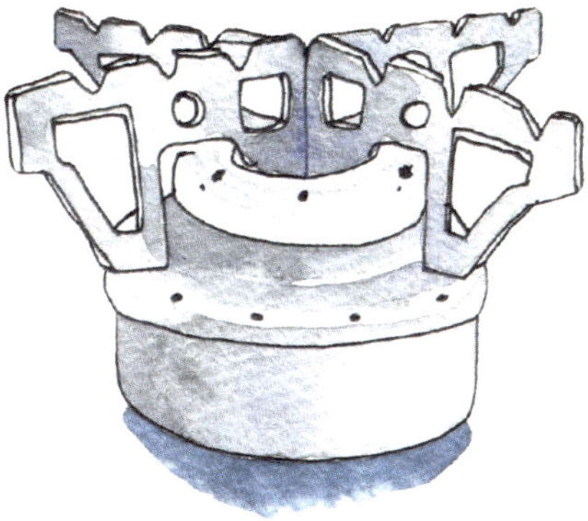

1. Select a well-ventilated outdoor area away from flammable materials, structures, and overhanging branches.
2. Place the alcohol stove on a stable, flat surface. Ensure it won't tip over during use.
3. If your stove has a built-in pot holder, make sure it is attached properly.
4. Following the manufacturer's instructions, carefully pour your fuel into the stove's fuel reservoir. Use a funnel if needed to avoid spills.
5. Using your ignition source, carefully ignite the fuel in the stove. If your stove has a priming area, you may need to ignite the fuel there first to create a stable flame.
6. Once you have finished cooking, allow the flame to burn out naturally. If the stove has a shut-off mechanism, use it to extinguish the flame.

TIPS

- Some fuel is packaged in a temporary plastic container. It's important to transfer this fuel to a hardier storage container for long-term use or for carrying in packs.
- If you are using the stove outdoors, a windscreen can help protect the flame, allowing for more efficient cooking.
- Choose lightweight, heat-conductive pots and pans, such as aluminum or titanium. This helps reduce cooking time and fuel consumption.

- Alcohol stoves may take longer to heat up compared to other types of stoves. Be patient and allow extra time for cooking, especially with larger meals.

CAUTIONS

- Stand back and ensure your hands are clear of the flame while lighting the stove.
- Avoid overfilling the fuel reservoir, as this can lead to spills when placing the pot on the stove.
- After cooking, allow the stove to cool completely before handling it or packing it away. Alcohol stoves can retain heat, and touching them prematurely can lead to burns.
- Be aware that alcohol flames can be difficult to see, especially in bright daylight, so make sure the stove is out before handling it.
- Store any leftover fuel in a safe, sealed fuel container away from heat sources.
- Ensure the alcohol stove and any remaining fuel are stored in a cool, dry place, away from children and pets. Follow the manufacturer's guidelines for storage.
- Make sure the fuel reservoir is cool before you refill it with alcohol, as a hot reservoir may spontaneously light and the flame may follow the liquid into your storage bottle.

EARTH OVENS

Earth ovens (also known as pit ovens or ground ovens) are one of the oldest known cooking methods in human history. At their most basic, they involve digging a hole in the ground, heating stones or coals inside it, and using the residual heat to cook food slowly, often over several hours. A more complex version involves making an oven out of clay. In this project, we are going to keep things relatively simple.

MATERIALS

- Natural clay or earth
- Dry firewood (hardwood)
- Cooking vessels (see Tips)
- Green leaves such as corn husks or banana leaves
- Heat-resistant gloves
- Fire extinguisher or water

TOOLS

- Shovel or spade
- Pickaxe (optional)

KEY ACTIONS

1. Select a safe, flat area away from flammable materials, structures, and overhanging branches. Ensure the ground is stable and suitable for digging.
2. Use the shovel or spade to dig a hole in the ground. The size and depth of the pit will depend on the amount of food you plan to cook. A typical pit is 2 to 3 feet deep and 3 to 4 feet wide. **(A)**
3. Shape the walls of the pit to be as even as possible, which helps with heat retention.
4. Place dry firewood in the center of the pit.
5. Light the fire and let it burn for thirty to sixty minutes, until you have a good bed of hot coals. This will provide consistent heat for cooking. **(B)**.
6. Wrap the food in leaves or place it in heat-safe pots. If using leaves, make sure they are clean and suitable for cooking.
7. Once the fire has burned down and you have a bed of hot coals, carefully remove any remaining larger pieces of wood. **(C)**

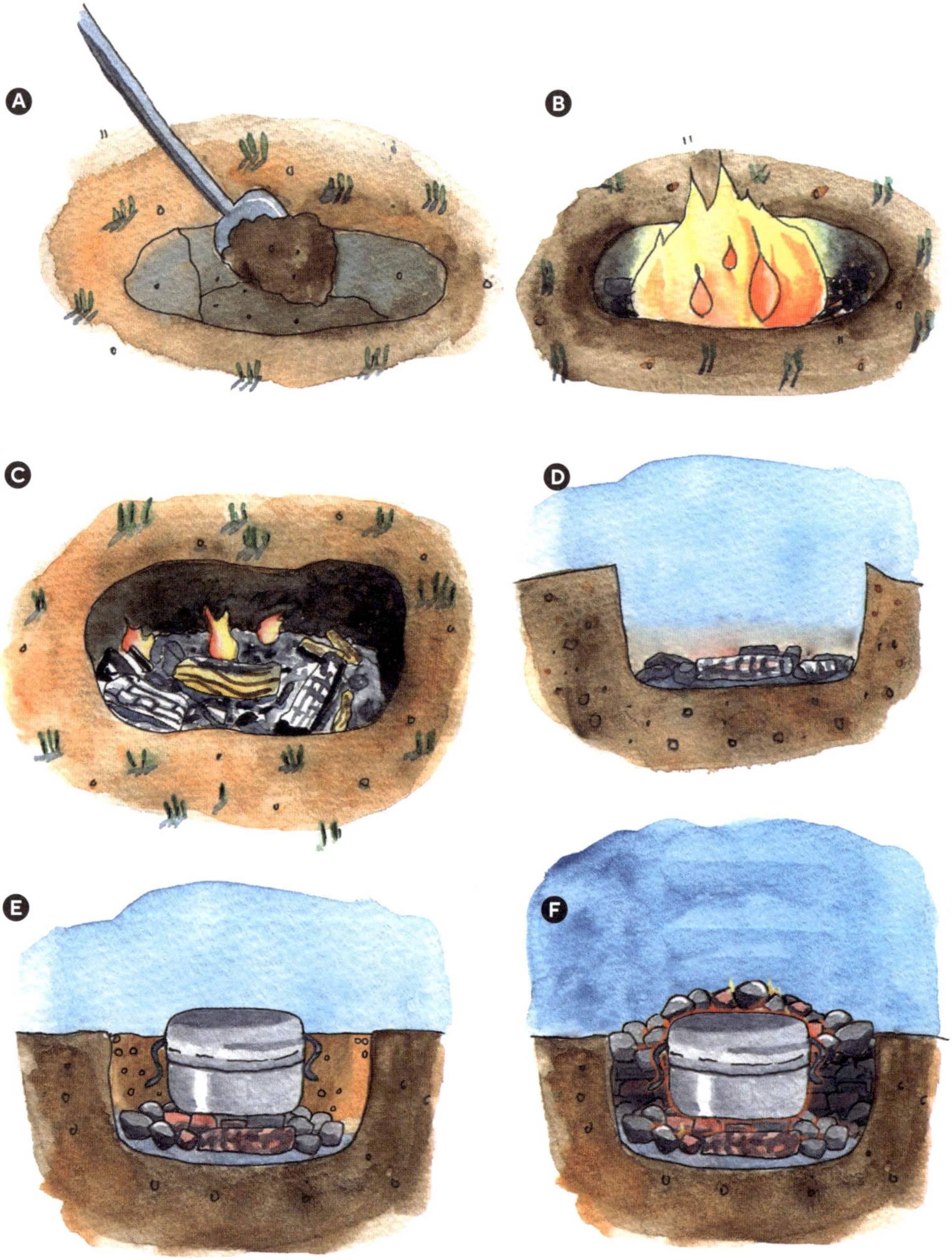
A
B
C
D
E
F

8. Create a hole in your coals to place your food into. **(D)**

9. Carefully place your wrapped food or pots into the hole and cover it with a layer of coals or soil, depending on the speed or heat you wish to cook with. **(E)**

10. Allow the food to cook for several hours, depending on what you are preparing. Cooking times can vary based on the type and size of the food. **(F)**

TIPS

- Use dry hardwood for fuel, as it burns hotter and longer, providing the necessary heat for cooking.
- Wrapping food in leaves (such as banana leaves or corn husks) or using cast-iron pots can help keep moisture in and protect the food from direct contact with the soil.
- Meats generally take longer to cook than vegetables, so plan your cooking schedule accordingly.
- Use heat-resistant gloves to remove the food from the pit.

CAUTIONS

- If using banana leaves or corn husks, ensure they are clean and safe for cooking.
- Avoid using treated wood or wood infused with chemicals, as it can release harmful toxins.
- When uncovering the food, be cautious of the steam escaping the pit.
- Allow the pit to cool completely before cleaning up.
- Keep a water source or fire extinguisher nearby for safety.

6 CREATE A WILDFIRE-DEFENSIBLE SPACE

In wildfire-prone areas, the landscape can be laden with dry vegetation, dead trees, and other combustible materials that can easily ignite. By establishing a defensible space, property owners create a buffer zone that minimizes the potential for flames to reach their homes. This not only protects the physical structure but also provides firefighters with a safer and more accessible area to work in the event of a wildfire. A well-maintained defensible space can significantly increase the chances of a home surviving a wildfire, as it reduces the intensity of the fire and slows its progression.

A cleared area for defensible space should generally extend at least 30 feet from your home and any structures. In more fire-prone areas, this distance may increase to 100 feet or more, depending on local regulations and the specific fire hazards in your region. This space should be free of combustible materials, including dead vegetation, debris, and flammable landscaping.

Creating a defensible space requires careful planning and ongoing maintenance. This thoughtful preparation not only safeguards homes but also can enhance the overall aesthetic appeal of the property, creating a more enjoyable outdoor environment.

WHEN TO DO IT

Work on creating a defensible wildfire space in the spring and fall, when vegetation is actively growing. This makes ground clearing and maintenance a lot easier. Regular upkeep during these seasons helps reduce fuel loads and enhances the safety of properties against potential wildfires. If you haven't done the maintenance in prior seasons, most definitely do the work heading into summer as everything dries out.

MATERIALS

- Mulch
- Fire-resistant plants
- Gravel or rock
- Fencing
- Water supply (hoses, water tanks, sprinklers)

TOOLS

- Pruner
- Rake
- Shovel
- Chainsaw
- Brush cutter
- Mower

KEY ACTIONS

1. Walk around your property to identify areas with dense vegetation, dead trees, or debris that could fuel a fire.
2. Measure and mark at least 30 feet around your home (if possible), extending up to 100 feet or more in high-risk areas.
3. Remove dead leaves, branches, and any flammable debris from the marked area. Rake up any loose material.
4. Prune trees and shrubs to create vertical and horizontal separation. Trim branches at least 6 to 10 feet above the ground and maintain a distance of 10 feet between tree canopies.
5. If possible, create a firebreak using gravel, rocks, or nonflammable materials to separate your property from surrounding vegetation.
6. Try to replace flammable landscaping with fire-resistant, drought-tolerant plants (see page 80). Space them adequately to prevent fire from spreading.
7. Keep grass mowed to a height of 3 inches or less and water regularly to keep it green.
8. Use fire-resistant materials (metal, concrete, tiles, or treated woods) for decks, fences, and outdoor furniture. Avoid using wood mulch near the home (see Tips).
9. Install water tanks with long hoses and emergency sprinkler systems to ensure that your house and surrounds can be adequately wet down in times of fire.
10. Clean out gutters.

FIRE-RESISTANT PLANTS

Replace flammable landscaping with fire-resistant, drought-tolerant plants. Space them adequately to prevent fire from spreading. Some fire-resistant plants include:

- **Lavender:** Aromatic and drought-tolerant, lavender is known for its resistance to flames. **(A)**
- **Sage:** Many varieties of sage are drought-tolerant and have a low resin content, making them less flammable. **(B)**
- **Yarrow:** This hardy perennial has a high moisture content and is less likely to ignite. **(C)**
- **Sedum:** Also known as stonecrop, sedums are succulent plants that retain moisture. **(D)**
- **Aloe Vera:** This succulent is not only useful for its medicinal properties but also has low flammability. **(E)**
- **Manzanita:** These evergreen shrubs are drought-resistant and have a lower risk of catching fire.
- **Bee Balm:** Known for attracting pollinators, bee balm has a high water content and is relatively fire-resistant. **(F)**
- **Ice Plant:** These ground covers are succulent and help retain moisture, reducing flammability. **(G)**
- **California Lilac:** Native to California, these shrubs are drought-tolerant and can help create a fire-resistant landscape.
- **Barberry:** Some varieties are more fire-resistant and can serve as effective hedges. **(H)**

A

B

C

D

E

F

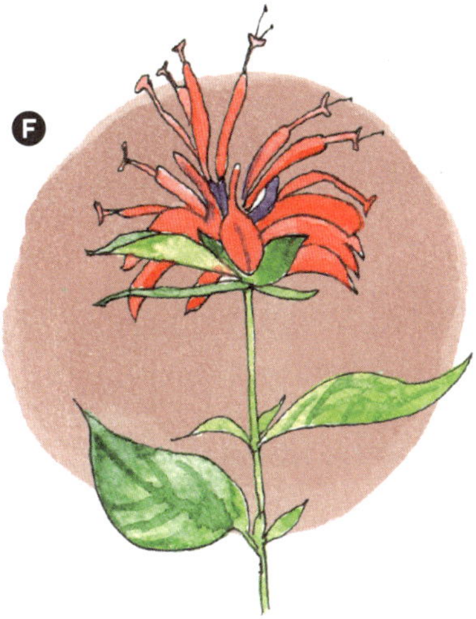

G

H

TIPS

- Use rubber mulch instead of wood mulch. It's nonflammable and durable, and it can add a fun pop of color to your garden while also being fire-resistant.
- If you compost, avoid adding highly flammable materials like dry leaves and twigs. Instead, focus on green waste to keep your compost heap moist and less flammable.
- Border garden beds with decorative stones or gravel. It not only looks nice but also creates a nonflammable zone that can slow the spread of fire.
- Set up a small, shaded area for pets with fire-resistant materials, ensuring they have a cool spot during hot weather while also being safe from any potential fires.

CAUTIONS

- Always consult your local fire department or forestry service for specific guidelines and regulations regarding defensible space and fire safety.
- Steer clear of flammable outdoor decorations, such as wooden furniture or fabric canopies, which can easily catch fire.
- Pay attention to local weather reports for fire danger ratings and red flag warnings, and be prepared to take action if conditions worsen.
- Don't overcrowd your defensible space with plants or decorative items, as this can create additional fire hazards.
- Always listen to emergency responders. Don't assume that it is safe to stay, even if you have done everything to prepare your house for fire. Your property can be replaced but lives cannot.

7 EARTHQUAKE-PROOF YOUR HOME INTERIOR

Earthquakes can occur suddenly and without warning, leading to significant structural damage and presenting serious risks to anyone inside. Small adjustments inside your home can make a big difference in creating a safer living environment for you and your family.

During an earthquake, unsecured items can become dangerous projectiles, posing serious risks of injury. By taking simple steps to secure furniture, appliances, and personal belongings, you greatly reduce the likelihood of accidents. Minor adjustments can save you from expensive repairs and replacements, ultimately minimizing the financial impact of an earthquake.

When families take proactive measures to protect their space, it fosters a sense of security and peace of mind, particularly for those living in earthquake-prone areas. Knowing that you have taken steps to safeguard your home can alleviate anxiety for you and your loved ones.

Many earthquake-proofing measures are simple and affordable, and small changes can often be implemented quickly and at a minimal cost, making them accessible to most homeowners.

Establishing an emergency plan is also crucial. Designating safe spots in your home where family members can take cover during an earthquake can make a significant difference in ensuring everyone knows what to do when the ground shakes. Practicing response drills reinforces this knowledge and helps everyone feel more confident and ready.

While large structural upgrades are important for earthquake safety, small adjustments within your home can have a profound impact. By focusing on securing items, creating safe spaces, and establishing emergency plans, you can significantly enhance your home's resilience against earthquakes.

WHEN TO DO IT

Earthquake-proof your home interior as soon as you move in, especially if you live in an earthquake-prone area. Next to then, now is a good time. Regularly reassess your home's preparedness, especially after renovations or when acquiring new furniture or appliances.

MATERIALS

- Heavy-duty straps or brackets to secure large furniture items, such as bookshelves and cabinets, to walls
- Adhesive putty
- Rubber mats
- Wood or metal braces for reinforcing doorways and windows
- Heavy-duty wall anchors for securing items to drywall or other wall materials
- Duct tape

TOOLS

- Drill
- Hammer
- Screwdriver
- Level

KEY ACTIONS

1. Identify the safe areas of your house to position yourself in during an earthquake. The best areas are under sturdy furniture or against an interior wall.
2. Walk through your home and identify items that could pose hazards during an earthquake. Look for heavy furniture, appliances, and fragile items that could fall or tip over.
3. Using anchoring straps or brackets, secure tall furniture such as bookshelves and cabinets to the wall. Locate the wall studs for proper anchoring and use a drill to install the straps according to the manufacturer's instructions. For heavy appliances such as refrigerators and water heaters, use anti-tip brackets or straps to secure them to the wall. This helps prevent items from tipping over during seismic activity.

4. Apply adhesive putty to the bases of decorative items, picture frames, and fragile objects on shelves. This will help keep them in place and reduce the risk of breakage during an earthquake.
5. Place rubber mats or nonslip pads under heavy furniture and appliances to prevent sliding. This will help keep items in place during tremors.
6. If necessary, use wood or metal braces to reinforce doorways and windows. This adds stability and helps maintain their structure during an earthquake.

TIPS

- Develop an emergency plan with your family. Identify safe spots in your home where you can take cover during an earthquake, and establish a communication plan in case you are separated.
- Designate a small area in your home as a "safe room" where family members can gather during an earthquake. Ensure it's free of heavy furniture and other hazards.
- Consider applying safety film to windows or using shatterproof glass to reduce the risk of glass shattering during an earthquake.
- Store heavy items on lower shelves and avoid overloading upper shelves with breakable or heavy objects. This reduces the risk of items falling during shaking.
- Periodically reassess your home's earthquake preparedness, especially after acquiring new furniture or making changes to your living space.
- Practice earthquake-response drills with your family so everyone knows what to do when the ground shakes. Rehearsing the plan will help everyone feel more prepared and confident.

IF DISASTER STRIKES

Where you are when the earthquake strikes will determine your course of action, but the main idea, no matter where you are, is to:

1. Stop: Do not try to keep moving or you may fall and get injured.
2. Drop: Get close to the ground for more stability.
3. Cover: Seek shelter under something solid or cover your head if you can't find anywhere to get under.
4. Hold on: Grab on to something solid if the ground is buckling.

As soon as the main tremors cease and debris stops moving around, check yourself and others for injuries. **If you live in coastal areas, make your way to high ground as soon as possible, as there is a chance a tsunami may follow.** Don't wait for authorities to advise you on this matter—by the time you realize a tsunami is on its way, it's too late.

Check water, gas, and power lines for damage. Stay out of affected areas of your house and avoid or clean up any broken glass and debris. Turn on the radio to see what authorities are advising and make sure you follow their instructions. Open cupboards carefully and filter and boil all tap water until you have been told it is safe to drink.

CAUTIONS

- During an earthquake, stay away from windows, glass, and exterior walls, as these areas are more likely to collapse or shatter.
- Do not shelter under door jams. This used to be the advice for a safe space in an earthquake, but houses are built differently these days and this is no longer a good place to position yourself.

8 STORM-PROOF YOUR HOUSE

Storm-proofing focuses on practical measures that enhance safety. For typical storms, high winds and heavy rainfall can lead to issues such as water leaks, debris damage, and flooding. Simple steps like ensuring that gutters and downspouts are clear can help redirect rainwater away from the foundation, reducing the risk of flooding and water intrusion. Additionally, securing outdoor items such as patio furniture or garden decorations can prevent them from becoming projectiles in strong winds.

When preparing for hurricanes or tornadoes, effective measures can be taken without extensive renovations. Installing storm shutters or using plywood to cover windows protects them from breaking due to high winds or flying debris. Reinforcing doors with additional locks can help them withstand powerful gusts. Ensuring that your roof is in good condition and that shingles are securely fastened can also prevent damage from wind uplift.

Maintaining a well-organized emergency kit (see page 284) and having a communication plan (see page 235) in place are other important aspects of storm-proofing. This allows your family to react quickly and safely when a storm approaches. Also, keeping important documents in a waterproof container ensures that they remain safe during potential flooding.

By focusing on these practical measures, you can significantly enhance your home's resilience to storms. These proactive steps not only protect your property but also provide peace of mind, knowing that you are better prepared to face the challenges that storms may bring.

WHEN TO DO IT

Storm-proof your house well in advance of storm season to ensure all preparations are complete before severe weather arrives. Conduct regular maintenance and checks throughout the year to address any new vulnerabilities that may arise.

More extreme hurricane or tornado measures like storm shutters can be prepared and stored somewhere accessible to be brought out when warnings are issued.

MATERIALS

Many of these materials are optional, depending on the severity of the storms in your area. I have provided a complete list so you can prioritize based on your needs.

- Storm shutters or plywood
- Sandbags
- Gutter guards
- Roofing repair materials
- Heavy-duty straps or anchors
- Weatherproofing caulk
- Buckets
- Plastic containers
- Emergency radio
- Nails and screws

TOOLS

- Hammer
- Drill
- Measuring tape
- Saw
- Ladder

ACTION ITEMS IN ADVANCE OF AN EMERGENCY

1. Research your local area for storm susceptibility. Are you in the tropics where hurricanes are prevalent? Does your area get regular tornadoes at certain times of year? Are you in a low-lying area that may experience flooding? The answers to these questions will dictate what your storm preparation priorities are and how best to prepare. Even if you aren't in an area that is regularly affected by extreme storms, it might be worth having a few supplies on hand for unprecedented events.
2. Walk around your home and identify areas that may be susceptible to wind, rain, or flooding. Look for loose tiles, rotten window seals, or loose shutters, for example. Pay close attention to windows, doors, and the roof. Make a plan to reinforce or repair those areas.
3. Remove any debris from gutters and downspouts to ensure proper drainage. This will help divert rainwater away from your foundation.

4. Double-check that all windows and doors are properly sealed to prevent water intrusion. Use weatherproofing caulk to seal any gaps around windows and doors.
5. Inspect your roof for missing or damaged shingles. Apply roofing cement or sealant to repair any issues and reinforce the integrity of your roof.
6. Purchase sandbags if you are in an area susceptible to flooding (see page 91).
7. Assemble an emergency kit with enough supplies to last three days. This should include nonperishable food, bottled water, a flashlight, batteries, a first-aid kit, and important documents stored in waterproof containers (see page 284).
8. Write up an action plan for when a storm is close. This will be the last-minute things you will need to do as the storm approaches (see page 89).
9. Discuss with your family the steps to take during a storm and establish a meeting point in case of evacuation.

ACTION ITEMS WHEN A STORM IS APPROACHING

When a storm is approaching, it's important to monitor local weather reports and alerts to keep track of the storm's progress. In addition:

1. Bring in or securely fasten patio furniture, decorations, and any other items that could become projectiles in strong winds.
2. If you have storm shutters, install them on windows and doors (see page 89). If using plywood, secure it firmly to protect glass from breaking. These shutters can be already cut to measure, pre-drilled, and stored in a safe, accessible place (see page 90).

FOR PERMANENT STORM SHUTTERS:

- Ensure you have all necessary materials, including storm shutters, screws, anchors (if applicable), and a screwdriver or drill.

- Inspect the storm shutters for any damage or wear. Make sure they are in good condition before installation.
- Position the shutters over windows and/or doors, ensuring they cover the entire opening.
- Use a pencil to mark the screw or anchor points on the wall or window frame where the shutters will be secured.
- If necessary, drill holes at the marked points, making sure to select the appropriate drill bit size for your anchors or screws.
- Align the shutters with the drilled holes and secure them in place using screws or anchors. Tighten them until the shutters feel secure but avoid overtightening to prevent damage.
- If your shutters have a folding or rolling mechanism, test it to ensure it opens and closes smoothly.

FOR TEMPORARY STORM SHUTTERS (PLYWOOD):

- Collect plywood sheets, a measuring tape, a saw (if needed), screws, and a screwdriver.
- Measure the dimensions of each window or door that needs protection.
- Cut the plywood to fit the measurements, ensuring it is slightly larger than the opening for adequate coverage.
- Pre-drill holes in the plywood for screws to make installation easier and prevent the wood from splitting.
- Hold the plywood against the window or door, ensuring it covers the entire area.
- Use screws to attach the plywood to the wall or window frame, placing screws every 12 to 16 inches along the edges for stability.
- Ensure that the plywood is securely fastened and does not wobble or move.

SANDBAGS: WHAT YOU NEED TO KNOW

Sandbags are a simple but effective way to reduce flooding and divert water. Here's what you need to know to obtain, fill, and place them for the best protection.

WHAT TO PURCHASE

Size:

- Standard 14 x 26-inch bags work best.

Material:

- Polypropylene (synthetic) sandbags are affordable, water-resistant, and last 6 to 12 months outdoors.
- Hessian/burlap sandbags are biodegradable and eco-friendly, and good for short-term use.

Pre-filled vs. empty sandbags:

- Pre-filled sandbags save time in emergencies.
- Empty bags are cheaper and store flat, but you'll need to have sand or soil at the ready.

Sand type:

- Brickie's sand or washed sand packs best. In a pinch, use soil.

HOW TO FILL

- Fill one-half to two-thirds (about 30 pounds).
- Overfilled bags won't stack or seal gaps properly.
- Don't tie the top; fold it under when stacking.

HOW TO PLACE/STACK

- Lay bags flat and lengthwise, overlapping them like bricks.
- Fold the open end under and face it against the water flow.
- For doorways or garage gaps: Create a single barrier two to three bags high.
- For perimeter walls: Stagger bags in a pyramid shape (wide base, narrower top).

OTHER TIPS

- Clear debris and level the ground before placing sandbags.
- Use plastic sheeting behind or under sandbags to boost water resistance.
- Wear gloves. Sandbags can cause friction burns.
- Don't wait for floodwaters to arrive. Set up early while the ground is dry and safe.
- After use, dispose of contaminated bags (mud/sewage) separately from clean ones.

3. If you live in a flood-prone area, fill sandbags and stack them around vulnerable entry points, such as doors and windows, to help divert water (see page 91).

4. Charge your electronic devices and have backup power sources ready, such as portable chargers or generators (see pages 20 to 34).

5. Locate your emergency kit and prepare alternate light sources (see page 284 and page 35).

6. Review your evacuation routes and have a plan in place for where you will go if you need to leave your home.

TIPS

- Throughout the year, check your storm-proofing measures and make any necessary updates or repairs to ensure your home remains prepared for future storms.
- In addition to having physical copies of important documents, consider scanning and storing them in a secure cloud service for easy access if you need to evacuate.
- Trim any overhanging branches and remove dead trees or shrubs that could fall and cause damage during high winds.
- Document your belongings, including photos and descriptions, to assist with insurance claims in case of damage.
- Keep your vehicle's gas tank full and ensure that it is maintained and ready for travel in case of evacuation.
- Check your insurance documents to make sure you are covered for relevant storm damage, especially if you live in an area with a history of storm devastation.

CAUTIONS

- If the storm is severe and evacuation is necessary, leave early to avoid traffic and ensure you reach a safe location.
- Once the storm begins, stay indoors, away from windows, and wait for official announcements indicating it is safe to go outside.

9 SET UP A SAFE ROOM

Setting up a safe room in your house is crucial for emergency preparedness, providing a designated space where you and your family can seek refuge during severe weather events or other emergencies. There are many reasons why a safe room is important, including:

1. **Protection from hazards:** A safe room offers protection from high winds, flying debris, and other dangers associated with extreme weather, such as tornadoes or hurricanes.
2. **Centralized safety:** Having a designated space helps everyone in your household know where to go quickly in an emergency, reducing panic and confusion.
3. **Increased security:** A safe room can provide a secure location in case of intruders or other security threats, giving peace of mind. Gathering everyone in one place ensures the best measures for defense rather than having attention and resources scattered.
4. **Protection from natural disasters:** A safe room can serve as a refuge during other emergencies, such as earthquakes or civil unrest.
5. **Comfort and reassurance:** Having everyone in one space allows for a feeling of security and companionship.

There are many variations of a safe room, ranging from a small room, such as a bathroom or closet, that is away from windows and outside-accessing doors to a reinforced basement, panic room, or even an outside bunker or shelter. Which you choose will be up to things such as available space, budget, resources, and the likelihood of requiring one.

I recommend choosing one where you can create a safe sleeping area in case your emergency stretches from hours into days. Designate a specific corner or section of your safe room as a sleeping area. Ensure it has comfortable bedding, such as mattresses or airbeds, along with extra blankets and pillows.

You will need to ensure the safe room is easily accessible to all family members, including children and those with mobility issues, and regularly check and update supplies in the safe room, including food, water, medications, and batteries for flashlights and radios.

If you don't have the space for a permanent safe room, make a plan for which area would work best for one and have a system in place that can quickly transform the chosen space into a safe room should the need arise.

Setting up a safe room, complete with a safe sleeping area, can greatly enhance your family's safety and preparedness during emergencies, providing a reliable refuge when it's needed most.

The safe rooms we are going to explore in this chapter include:

- Basic safe space
- Reinforced room
- Panic room
- Family shelter

WHEN TO DO IT

Set up a safe room in your emergency preparedness plan as soon as you identify potential risks in your area, such as severe weather, natural disasters, or security threats. Establishing the safe room early ensures that your family knows where to go during an emergency and has adequate time to equip it with essential supplies.

BASIC SAFE SPACE

A basic safe space simply refers to a room with a solid door that can be securely closed and locked.

MATERIALS

- Basic emergency kit (see page 284 to decide your needs)
- Communication devices
- Blankets or sleeping bags and pillows
- A small battery-operated fan or air-circulation device
- Trash bags, wipes, and any necessary personal hygiene items
- Personal medications

KEY ACTIONS

1. Select a small, interior room away from windows and external doors, such as a bathroom, closet, or basement.
2. Ensure the door can be securely closed and locked. Consider adding a door wedge or barricade for extra security.
3. Stock room with essential emergency kit items (see page 284).
4. Place a charged mobile phone or two-way radio in the safe space for communication during an emergency.
5. Prepare for a sleeping area. Include blankets, sleeping bags, and pillows to ensure comfort if you need to stay in the room for an extended period. If possible, add a small battery-operated fan to improve air circulation and keep the space comfortable.
6. Stock personal medications and sanitation supplies such as toilet paper and hand soap.
7. Discuss the safe space with your family, ensuring everyone knows its location and understands when to use it.

REINFORCED ROOM

A reinforced room is a room with a solid-core or metal door that can withstand force.

MATERIALS

- Heavy-duty locks
- Metal brackets, doorjamb reinforcements, and strike plates to strengthen the doorframe
- Plywood sheets or drywall with additional studs to enhance wall strength
- Steel plates or ballistic panels for added protection against projectiles
- Shatter-resistant films or metal bars for window security; or consider eliminating windows, if possible

- Screws or nails
- Emergency kit (see page 284)
- A charged mobile phone or two-way radio for emergency communication
- Battery-operated fan or air purifier
- Blankets, sleeping bags, and pillows
- Trash bags, wipes, and personal hygiene items for cleanliness

TOOLS

- Drill
- Saw
- Hammer
- Measuring tape
- Wrench or screwdriver set
- Utility knife or multi-tool
- Ladder or step stool

KEY ACTIONS

1. Select an interior room that is away from windows and external doors. Ideally, it should be a small space, such as a basement, closet, or bathroom.
2. Replace any hollow-core doors with a solid-core or metal door.
3. Install a high-quality dead bolt and an additional locking mechanism for enhanced security.
4. Use metal brackets and strike plates to strengthen the doorframe against forced entry.
5. Reinforce walls with plywood sheets or steel plates to increase their strength. Use screws or bolts to secure them firmly.
6. If possible, add additional studs to the walls to enhance structural integrity.
7. If you are in a room with windows, apply a shatter-resistant film to the windows to prevent breakage. Consider installing metal bars for added security.
8. Stock room with emergency supplies.

9. Place a charged mobile phone or two-way radio in the safe room for emergency communication.
10. Install a battery-operated fan or air-purifying system to maintain airflow in the room. A portable air purifier is often sufficient, but if you decide that you want an air-filtration system, it is best to consult a professional.
11. Prepare a sleeping area. Include blankets, sleeping bags, and pillows to ensure comfort during extended stays.
12. Include trash bags, wipes, and personal hygiene items to maintain cleanliness.

PANIC ROOM

While a safe room and a panic room seem similar, a few small differences will help you to decide which is a priority for you. A secure safe room is designed for safety during emergency scenarios like natural disasters or home invasions, focusing on comfort and accessibility for prolonged use. It typically includes basic safety features such as reinforced doors, ventilation, and emergency supplies. In contrast, a panic room is specifically fortified for immediate protection against violent intruders, emphasizing security with reinforced walls and a solid-core or steel door with heavy-duty locks.

MATERIALS

- High-security deadbolts, electronic locks, or biometric locks for enhanced security
- Steel or ballistic panels to strengthen walls against attacks
- Security cameras and motion sensors to monitor surrounding areas
- A dedicated landline or mobile phone, plus a two-way radio for emergency contact
- A battery-operated fan, air purifier, or air-filtration system to maintain airflow
- Concealed storage for valuables or weapons

KEY ACTIONS

1. Select a suitable space. Choose an interior room, basement, or closet that can be easily secured and is away from windows and outside access points.
2. Remove the existing door and replace it with a solid-core or steel door.
3. Install high-security deadbolts, electronic locks, or biometric locks for added security.
4. Add plywood or ballistic panels to the walls, securing them with screws or bolts for added strength.
5. If possible, insulate the walls for soundproofing.
6. If your room has windows, apply shatter-resistant film to make them less vulnerable and consider installing security bars or grills.
7. Set up cameras outside the panic room to monitor activity.
8. Use motion detectors to alert you of any intrusions. This will allow you time to get into your panic room.
9. Place a dedicated landline phone or charged mobile phone inside the room, and consider adding a two-way radio for emergency communication.
10. Ensure adequate airflow. You might want to add a battery-operated fan or air purifier.
11. Practice getting to your panic room at night in the dark.

FAMILY SHELTER

A family shelter safe space is specifically designed to provide a secure environment during emergencies, whether located above ground, underground, or outside, such as in a fortified shed or bunker in the yard. Unlike other safe spaces, which may focus solely on individual safety or short-term refuge, a family shelter prioritizes the collective safety and comfort of all family members for a long-term scenario, often incorporating shared supplies and communal living elements. This type of shelter is equipped with features that cater to the needs of families, such as enough space for everyone, emergency supplies for all ages, and provisions for maintaining communication and comfort during prolonged stays.

MATERIALS

This section is going to assume that you already have the space constructed. Cellars, sheds, and underground bunkers require specialist building knowledge, and I highly recommend engaging a professional to design and build these structures.

Some things to ensure your structure has are:

- Sturdy walls made of concrete blocks or bricks
- Wood or steel framing for additional support and structure
- Reinforced roof panels
- A solid-core or reinforced door with secure locks
- If including windows, ensure they are shatterproof and/or protected with bars
- Good insulation materials such as foam board, fiberglass, or other natural insulating materials to regulate temperature
- Battery-operated or manual ventilation systems to ensure airflow

Other materials:

- Emergency kit (see page 284)
- Blankets, sleeping bags, and pillows
- Portable seating solutions for comfort
- Two-way radios or a landline phone
- Portable toilet or bucket
- Sanitation items
- Trash bags and wipes
- Fire extinguisher

- Smoke detector and carbon dioxide (CO_2) monitor
- Heating or cooling methods, such as battery-operated fans or propane heaters
- Entertainment, such as games, puzzles, and music
- Comfort items as required by individuals

KEY ACTIONS

1. Assess your needs. Evaluate how many family members will use the shelter and what supplies will be necessary. Make lists.
2. Decide on key components, such as ventilation, comfort items, sanitation items, and emergency supplies.
3. Research the prices of each item on your checklist and create a budget for the total expenses. If your budget is limited, prioritize essential items, such as structural materials and emergency supplies.
4. Acquire supplies.
5. Once all items are gathered, check them against your checklist to ensure nothing is missing.
6. Store all materials in the designated family shelter area, keeping emergency supplies easily accessible.

TIPS

- Involve your household in the planning and organizing. This means they will be able to customize to suit their needs and be familiar with the area if they need to utilize it.
- Include a deck of cards, board games, or a book to help pass the time and keep spirits up during prolonged stays.
- Add some way of washing yourself to the kit, such as a basin and washcloth with some soap. There are ways to wash that don't take much water. Keeping up with hygiene can help to raise spirits.

- Add a favorite blanket or stuffed animal for emotional comfort, especially for children.
- Keep a portable phone charger in the safe space to ensure you can keep your devices powered during emergencies.
- Pack fun snacks, such as trail mix, candy bars, or popcorn, to make the experience more enjoyable.
- Include basic art supplies, such as coloring books, crayons, or journals, to engage creativity and reduce stress.
- Add a battery-operated lantern or fairy lights to create a cozy ambiance.
- Consider including a small portable speaker or music player with pre-loaded playlists to help lighten the mood.

CAUTIONS

- Ensure the safe space has adequate ventilation to prevent suffocation, especially if you're using it for an extended period. If the room has no windows, consider using a battery-operated fan to circulate air.
- Avoid overcrowding the safe space with too many supplies, which can limit movement and create a feeling of confinement. Maintain enough space for everyone to sit or lie down comfortably.
- Avoid using candles or other open flames in the safe room. Instead, rely on battery-operated lights or flashlights to minimize fire risk.
- If children will be using the safe space, ensure any potentially dangerous items (e.g., small batteries, sharp tools) are kept out of reach.

10 STASH EMERGENCY CASH

Stashing emergency cash and precious metals is a crucial aspect of emergency preparedness that can significantly enhance your resilience during unexpected crises. In times of disaster, traditional banking systems may become inaccessible due to power outages, infrastructure damage, or widespread panic. Having a reserve of cash on hand ensures that you can purchase food, water, and other necessities even when electronic payment systems fail. This immediate access to cash can be vital for securing supplies from local vendors or participating in barter exchanges, where cash might be preferred over credit or debit transactions.

In addition to cash, investing in precious metals like gold and silver can serve as a hedge against economic instability. These metals are often sought-after during crises, making them a reliable form of wealth preservation. In a scenario where currency loses its purchasing power, having metals can provide a means of trade or investment, allowing you to maintain financial security when conventional assets may falter.

Having emergency cash and precious metals is not just about individual preparedness; it empowers you to navigate uncertain circumstances with confidence, providing a buffer against the unpredictability of emergencies while fostering a sense of security for yourself and your loved ones.

WHEN TO DO IT

Stock up on emergency cash and metals in good times before natural disasters, economic instability, or supply chain issues arise. It's also a good idea to keep an eye on prices for precious metals and buy when they are favorable.

MATERIALS

- Cash (small denominations for easy transactions)
- Precious metals (gold and silver coins or bars for value retention)
- A secure safe or lockbox to keep cash and metals safe

KEY ACTIONS

1. Determine how much cash and how many precious metals you want based on your budget and potential emergency scenarios.
2. Learn about different types of precious metals (gold, silver, platinum) and their market value. Decide which ones you want to invest in.
3. Allocate funds for both cash reserves and purchasing precious metals. Ensure a balance that meets your emergency preparedness goals.
4. Find a secure location, such as a safe or lockbox, to store your cash and precious metals. Consider fireproof and waterproof options for added protection.
5. Withdraw small denominations of cash from your bank. Aim for $20s, $10s, and $5s for easier use in emergencies.
6. Buy precious metals from reputable dealers or online platforms. Ensure you receive proper documentation for your purchases.

TIPS

- Check your emergency cash and metal reserves regularly, especially after significant life changes or market fluctuations, and replenish as needed.
- Consider investing in a mix of coins and bars, as coins may be easier to trade for goods and services in an emergency due to their smaller denominations.
- Bury or hide a small amount of cash and metals in a discreet location on your property for a truly secure backup. Avoid obvious spots, and consider keeping a secure map or coded reminder.
- Follow market trends and news related to precious metals; sometimes, seasonal changes or global events can lead to favorable buying opportunities.

- Think about adding items that can hold value, such as collectible coins, vintage jewelry, or even tools and supplies that could be bartered in emergencies.

CAUTIONS

- Storing cash and precious metals at home can pose a theft risk. Ensure your storage is secure and consider using a safe or a bank safety deposit box.
- The value of precious metals can fluctuate significantly. Stay informed about market trends to avoid buying at peak prices or selling in a downturn.
- Be cautious when purchasing precious metals from unknown sources. Always buy from reputable dealers and verify authenticity.
- Ensure that your emergency cash and metals are easily accessible in a crisis. If they are difficult to access or retrieve quickly, they may not provide the necessary support.
- Cash can lose value over time due to inflation. Balance your cash reserves with investments in precious metals or other assets that can retain value.
- Ensure that family members know where cash and metals are stored. Lack of communication can lead to confusion during emergencies.

11 BUILD UP SUPPLY CACHES

Supply caches are essential components of emergency preparedness, offering hidden reserves of crucial items in both familiar and remote locations. When situated close to home, these caches provide immediate access to essential supplies during emergencies such as natural disasters or civil unrest. This proximity allows you to quickly retrieve necessary items without the need to venture out, which can be particularly beneficial in chaotic situations where public panic or looting may occur.

Having a supply cache nearby also serves as a backup resource. If your primary supplies become depleted or damaged, you can rely on the nearby cache to ensure you have food, water, medical supplies, and other necessities readily available. This setup not only enhances your security but also fosters a sense of community, as local caches can be shared among trusted neighbors, promoting cooperation and resource-sharing during times of need.

When it comes to survival, you want to ensure you are covering your basic needs, which are shelter, water, safety, and food. Although it might seem like a lot, these needs can all be met with supplies packed into a small daypack.

You can add your own flair and personal touches to these kits. If you prefer a multi-tool rather than a fixed-blade knife or think a pack of cards would make for a great morale booster, pop these things in. Just remember that a cache is best being as small as possible while still covering all your needs, so keep it to needs rather than too many luxury items.

Supply caches play a vital role in enhancing your preparedness for unforeseen circumstances, offering both immediate support close to home and strategic resources at bug-out locations. They will empower you to navigate crises with greater flexibility, security, and peace of mind.

WHEN TO DO IT

Create emergency caches well in advance of potential emergencies or natural disasters. This allows you to evaluate and replenish the caches as needed.

MATERIALS

- Waterproof and weather-resistant bins or bags
- Nonperishable food (three-day supply for all family members)
- Water bottles
- First-aid supplies and spare medication
- Multi-tool/knife
- Flashlight and extra batteries
- Fire-starting methods
- Extra warmth layers (thermal underwear)
- Emergency blankets
- Small tarp for emergency shelter
- LifeStraw and water purification tablets
- Portable charger with cords
- Small emergency radio
- Cash or barter items such as expensive watches or jewelry

KEY ACTIONS

1. Assess your needs. Determine the purpose of the cache (emergency supplies, food, tools, etc.) and consider how many people it needs to serve and the duration of support required. I recommend a three-day supply minimum in each cache.

2. Select a discreet but accessible location that is safe from weather elements and potential theft. Ensure it's a place you can easily return to when needed. Some locations to consider are:

    - Hide it beneath a pile of rocks or inside a hollowed-out tree stump.
    - Bury it under a thick layer of leaves or brush to blend in with the surroundings.
    - Use natural features like boulders or dense foliage for concealment.
    - Stash it inside an abandoned shed or barn, ensuring it's well-hidden (if you have access).

PART 2

WATER

WATER IS LIFE. We can survive less than three days without it. It's a fundamental resource that we often take for granted, especially in the hustle and bustle of urban living. Yet, when disaster strikes—be it a natural calamity, a man-made crisis, or even a simple infrastructure failure—having a reliable water supply becomes paramount. Imagine waking up to a citywide electricity outage that results in your taps not pumping water or facing a situation where your usual sources are compromised. It sounds daunting, and it can be, but with the right preparation, you can navigate these challenges with confidence.

This section will guide you through the essentials of ensuring that you and your loved ones have access to clean, safe water under any circumstance. We'll explore practical strategies for water storage, filtration, and purification, along with innovative methods to collect rainwater and use greywater effectively. You'll learn how to assess your personal needs, the needs of your family, and even your community's resources. Being prepared isn't just about stocking up; it's about understanding the systems around you and thinking ahead.

With a little foresight and planning, you can turn a potential crisis into an opportunity for resilience and self-sufficiency.

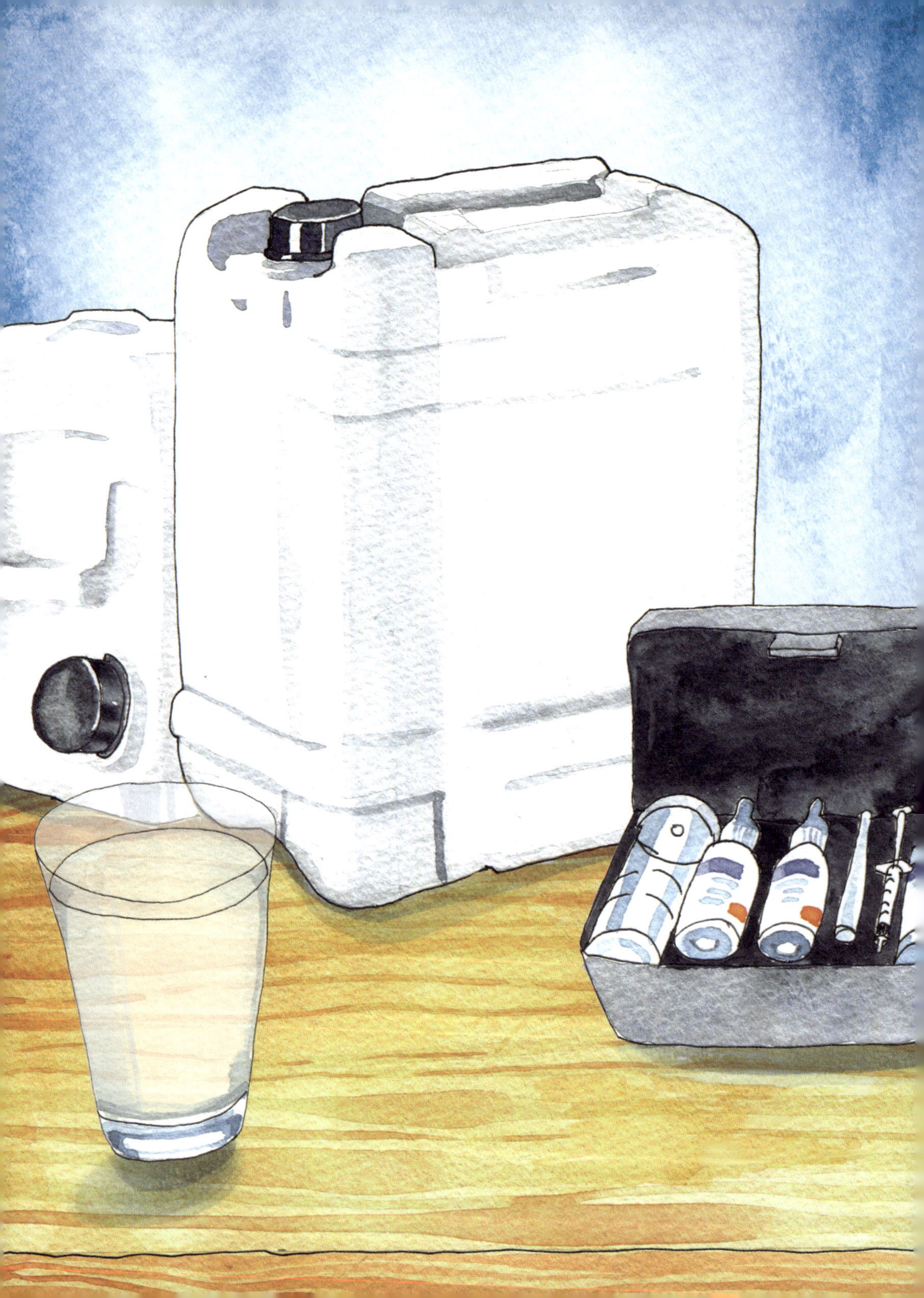

PROJECTS

12 ASSESS YOUR WATER NEEDS

Assessing your household's daily water usage is crucial for understanding consumption patterns and identifying areas for improvement. By utilizing the following methods, you can gain a comprehensive understanding of your household's water usage, which will empower you to make informed decisions about how much water you will need to store or access in an emergency. It will also allow you to start looking at where you can cut back on water usage and waste if you live in areas of drought or lack.

MATERIALS

- Log or journal
- Timer
- Calculator
- Spreadsheet

KEY ACTIONS

1. **Monitor your water meter:** Locate the meter, usually found near the street or in your basement. Take an initial reading and then record the reading again after a twenty-four-hour period without using water. The difference will give you the total water consumed in a day. You can use this as a baseline for your monitoring. For a week, record your water meter reading at the same time each day. Subtract the previous day's reading from the current day's to find out how much water you used each day.

2. **Calculate by fixture:** Use a timer during activities (like showering or washing dishes) to track how long each activity takes.

   Toilets: Determine how many flushes your household uses daily. A standard toilet uses about 1.6 gallons per flush. Multiply the number of flushes by the gallons used.

   Showers: Measure the flow rate of your showerhead (typically 2 to 3 gallons per minute). Multiply the flow rate by the average duration of showers taken daily.

Faucets: Measure the flow rate of kitchen and bathroom faucets (around 2 gallons per minute) and estimate the time they are used daily for washing hands, dishes, and so on.

3. **Calculate by appliance:** Check the specifications of your taps and appliances (e.g., washing machines, dishwashers) to determine their water usage per load. Record how often you use them and calculate total water use over time. Record the number of cycles run per day to calculate total water use.

4. **Measure outdoor water use:** Measure how much water you use for outdoor activities, such as watering gardens or lawns. Measure the output of your hose or sprinkler in gallons per minute, then estimate how long you water each day.

5. **Evaluate outside water use:** The average flow rate for a garden hose is typically 5 to 10 gallons per minute (GPM), depending on the hose diameter and water pressure. For garden sprinklers, the flow rate can vary widely based on the type of sprinkler:
   - Oscillating sprinklers: Usually use about 1 to 3 GPM.
   - Stationary sprinklers: Generally use around 1 to 4 GPM.
   - Rotary sprinklers: Can use between 2 to 6 GPM.
   - Soaker hoses: Typically deliver about 0.5 to 2 GPM.

   It's important to check the specifications of your specific hose or sprinkler for more accurate measurements.

6. **Track drinking and cooking water:** Keep a log of how much water you use for cooking and drinking. This may include water used for boiling, rinsing fruits and vegetables, and making beverages.

7. **Monitor over time:** Keep a daily log for at least a week, noting the estimated water usage for each of the categories above. Analyze the data to identify patterns and see where you can reduce consumption.

8. **Use smart meters:** Consider installing a smart water meter that tracks water usage in real time and provides insights through a smartphone app.

9. **Sum it up:** At the end of each week or month, analyze the data you've collected. Look for patterns in your water usage and identify which fixtures or activities consume the most water.

TIPS

- Categorize your water usage by fixture (e.g., toilets, showers, faucets, appliances) to identify which areas contribute the most to your overall consumption.
- Calculate the percentage of total water usage attributed to each fixture or activity. This can provide a clearer picture of where most of your water is being used.
- Analyze how your water usage varies by season. For instance, you may notice increased usage during certain months due to activities like washing more laundry or using additional appliances.
- If available, use smart water meters or apps that track water usage in real time. These tools can provide detailed insights and help identify patterns more easily.

PRIORITIZE YOUR NEEDS OVER YOUR WANTS

In an emergency, it's essential to distinguish between the water you need to survive and the water that you typically use in everyday life.

The recommended daily allowance is 1 gallon per person, which covers drinking, basic cooking, and minimal hygiene. Ask yourself: How does my current water usage compare to this? What everyday water uses might not be necessary in a life-or-death situation?

Although bathing might be nice, hygiene in an emergency should focus on preventing illness, such as washing hands after using the bathroom, before meal prep, or after handling chemicals. A hot bath or shower can definitely wait until the crisis has passed.

Consider water-saving alternatives such as sanitary wipes or hand sanitizer to conserve your limited supply.

13 STORE EMERGENCY WATER IN CONTAINERS

We don't tend to be aware of the amount of times we turn on a tap for convenience until it is gone. Practice observing your water use and seeing where you can cut back. Think about other ways of achieving the same goal without an excess of water. Simply anticipating scarcity will prepare you if the pumps do fail.

Now that you know how much water you use (see page 114), and that you should allow for 1 gallon per day per person in emergency situations, let's look at how to store this emergency water source. Most urban disaster scenarios will become manageable after three days, so it is recommended that you have at least three days' supply of emergency water in your house. I'll let you do the math.

Keep in mind that water may still be coming out of the taps, but it may not be safe to consume or use. Quite often natural disasters will cause pipes to leak, and sewage and other contaminants can end up coming out of your taps instead of the clean water we are used to. If you have had a natural disaster in your area, monitor what the government advisory is about your water supply and, if in doubt, break open your emergency supply.

WHEN TO DO IT

Organize your emergency water storage well in advance of potential emergencies, ideally before hurricane season or flu season. Regularly check and refresh your supply every six months to a year.

MATERIALS

- Food-grade water storage containers, such as plastic jugs or barrels
- Unscented household bleach
- Measuring spoon
- Permanent marker
- Water test kit (optional)
- Water purification tablets (optional)
- Funnel
- Labels

KEY ACTIONS

1. Choose a cool, dark, and dry place, away from direct sunlight and heat sources, to store your water containers (exposure can cause algae growth and degrade the container material). Basements or closets are ideal.

2. Wash the containers with soap and water, then rinse thoroughly. For extra sanitation, you can rinse the containers with a solution of 1 tablespoon of unscented bleach per gallon of water.

3. Fill the containers with clean, drinkable water (a funnel is helpful for avoiding spills). If using tap water, let it run for a few minutes before filling to cleanse the pipes of any contaminants. For long-term storage (three weeks or more), consider adding water purification tablets according to the manufacturer's instructions. If contaminants have entered the water, the water can develop bacteria, viruses, or algae. The tablets also allow the water to remain potable for years rather than months. Make sure the containers are tightly sealed to prevent contamination.

4. Label the containers with the date the water was stored as well as any additional information, such as "water from a tap," if you have stored tap water for cooking, sanitary, and hygiene purposes as well. This helps you keep track of freshness.

5. Place the sealed, labeled containers in your designated storage area. Add a reminder to your calendar every six months to inspect and possibly rotate your water. Check for leaks, discoloration, or cloudiness.

TIPS

- Opt for containers that are free from BPA (bisphenol A) to avoid chemical leaching into the water.
- Consider smaller containers for easier handling and to reduce waste if you need to use some of the water.

- Every six months, check the containers for leaks or signs of contamination. Rotate the water supply, replacing it with fresh water as needed.
- Document your storage locations, amounts, and expiration dates to easily track your water supply.
- If in doubt of the quality of your water, bring it to a rolling boil, pass it through a factory-bought water purifier, or use water purification tablets before consuming.

CAUTIONS

- Don't reuse milk or juice jugs. Even after these containers are cleaned, remaining sugars and proteins can promote bacterial growth.
- Avoid storing water directly on concrete floors. Certain types of concrete can leach chemicals over time. Use a pallet or liner as a barrier.
- Avoid storing water near chemicals. Keep your water stores away from cleaning supplies, fertilizers, and other chemicals to prevent contamination.
- Ensure that containers are thoroughly cleaned and sanitized before filling them with water to prevent bacterial growth and contamination.
- Avoid using containers that previously held non-food substances, as they may contain harmful residues.
- Regularly check your water storage containers for leaks or signs of deterioration, as this can compromise the safety of the water.
- If the water develops an unusual odor, color, or taste, do not consume it, as it may be contaminated.

14 INSTALL A RAINWATER HARVESTING BARREL

A rainwater harvesting barrel is a food-grade container, typically placed at the base of a downspout, that collects and stores rainwater runoff from a roof. Installing these barrels is a crucial step in preparedness, especially in the face of increasing water scarcity and extreme weather events. These barrels provide a sustainable and eco-friendly method of collecting and storing rainwater for various uses, such as irrigation, washing, and even potable water (with proper filtration). By capturing rainwater, you can reduce your dependency on municipal water supplies, which may become unreliable during droughts or emergencies. Note that in order to install a rainwater harvesting barrel, your house must be equipped with a well-functioning gutter system to direct rainwater into the barrel.

WHEN TO DO IT

Install rainwater harvesting barrels in the spring or fall. In spring, you can take advantage of the increased rainfall that often comes with the season. Fall is also ideal, as it allows you to prepare for winter and ensure your system is ready to capture both rain and melting snow. Installing barrels during these seasons also gives you time to assess your system's efficiency before the peak rainy season, helping you stay well prepared for wet weather.

MATERIALS

- Rainwater harvesting barrel(s) (made of food-grade plastic or other nontoxic materials, 50 to 100 gallons in capacity), with manufacturer's instructions
- Downspout diverter
- Screen or filter
- Spigot or faucet (if not already attached to barrel)
- Overflow valve
- Support structure (wooden base or raised platform to assist gravity feed; optional)
- Sealant or caulk
- Teflon/plumber's tape

TOOLS

- Drill and appropriate drill bits
- Saw or hacksaw
- Screwdriver or wrench

KEY ACTIONS

1. Select a suitable location for your rainwater harvesting barrel, ideally near downspouts to minimize the distance water needs to travel. Make sure the location is easily accessible for maintenance and water collection. The ground should be level.
2. If your water barrel doesn't already have a tap, drill a hole near the bottom of the barrel and install the spigot or faucet. Use Teflon or plumber's tape on the spigot threads to ensure a watertight seal. Use sealant or caulk around the edges to prevent leaks.
3. Clean the rainwater barrel thoroughly. If it has a lid, ensure it fits securely to prevent debris and insects from entering.
4. Place the barrel on a stable platform or stand to elevate it, allowing for gravity-fed access to the spigot. Ensure it is level and secure.
5. Cut the downspout at a height that allows for a diverter to be installed. Follow the manufacturer's instructions to attach the diverter, ensuring it directs water into the barrel while allowing overflow to continue down the original downspout. Ensure the overflow path directs water away from your home's foundation.

6. Place a screen or filter over the top opening of the barrel to keep out debris and insects. Ensure it is securely attached.
7. If your barrel doesn't have a built-in overflow valve, install one near the top. This will direct excess water away from the barrel when it's full, preventing overflow. This can be as simple as a pipe that funnels the water down a storm drain and away from the property so flooding doesn't occur.

TIPS

- Check local regulations and guidelines regarding rainwater harvesting.
- Test the system by waiting for rain or using a water hose. Check for leaks and ensure water flows into the barrel and through the spigot.
- If you plan to use the collected water for drinking, consider acquiring additional filtration and purification systems. Use a chemical testing kit to test the quality of the water. Label non-potable water barrels clearly if water is not safe for drinking.
- Install mosquito-proof mesh (1 mm or finer) to prevent breeding.
- Regularly check and clean the gutters, diverter, and screens to ensure optimal performance. Monitor the barrel for any signs of wear or damage and clean it periodically to prevent algae growth.

CAUTIONS

- Regularly inspect and clean your gutters, downspouts, and screens to prevent debris buildup, which can block water flow and lead to overflow.
- Make sure the barrel is placed on a stable and level surface. If elevated, the support structure must be strong enough to hold the weight of a full barrel.
- In colder climates, be mindful of freezing temperatures, which can damage the barrel and plumbing. Consider draining the system before winter.
- Regularly check for signs of pests, such as rodents or insects, and take preventive measures to keep them away from the system.

15 BUILD A DIY WATER FILTER (TWO-BUCKET SYSTEM)

A DIY water filter using a two-bucket system is a straightforward and effective method for cleaning water that utilizes readily available materials and is quite simple to set up. *This method does not make the water safe to drink.* It is not a purifier—it simply removes larger particles to make the water cleaner-looking.

This filter system consists of two buckets, where one holds unfiltered water and the other collects the filtered water. The top bucket is filled with various filtration materials, such as gravel, sand, and activated charcoal, which work together to remove impurities, sediments, and contaminants as water flows through them.

Using a two-bucket water filter is advantageous for several reasons: (1) It is cost-effective. (2) It utilizes basic materials. (3) It is simple to construct and maintain. The system significantly improves water quality by reducing sediments, odors, and chemical contaminants, thereby making the water safer for non-potable uses such as gardening or cleaning. In times of emergencies, this filter provides a reliable method for purification, making it a practical solution for improving water quality.

WHEN TO DO IT

This system is intended for use after an emergency or disaster has occurred. However, it is essential to ensure you have the basic materials ready and accessible in advance.

MATERIALS

- 2 x 5-gallon food-grade buckets
- 1 inch (3 pounds) coarse gravel to form the bottom layer
- 3 inches (6 pounds) clean sand to trap smaller particles (use fine sand for better filtration)
- 4 inches (6 pounds) activated charcoal
- Cotton cloth or coffee filters

TOOLS

- Drill or other sharp tool

KEY ACTIONS

1. Prepare top bucket: Drill a small hole (about 1/4 inch in diameter) in the bottom. This hole will allow the filtered water to flow into the second bucket.
2. Clean and sanitize both buckets using a diluted bleach solution (1 tablespoon of unscented bleach per gallon of water), then rinse thoroughly.
3. Place a layer of cotton cloth or a coffee filter on top of the charcoal layer to catch any remaining fine particles.
4. Add a layer of activated charcoal, 1 to 2 inches thick. This layer will remove chemicals and odors and will improve the taste of the water. **(A)**

A

B

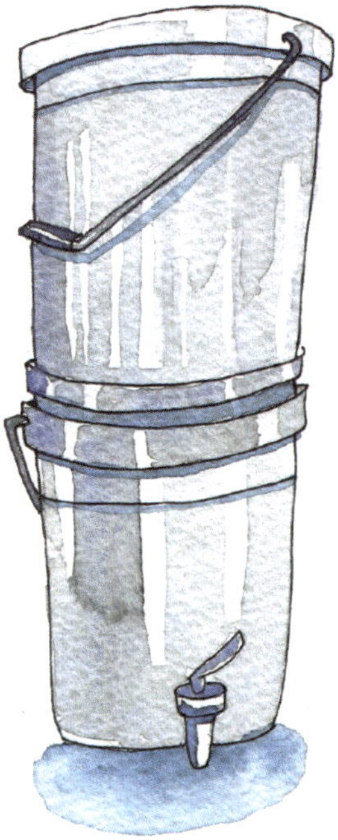

5. On top of the charcoal, add a layer of sand, 2 to 3 inches thick. This layer will help filter out smaller particles. **(A)**

6. Fill the remainder of the top bucket with a layer of gravel, 2 to 3 inches thick. This layer will catch larger particles and debris. **(A)**

7. Arrange the top bucket (with the filter material) above the bottom bucket. This can be done by hanging or placing the upper bucket on a step. Ensure that the drainage hole is aligned with the second bucket so that filtered water can flow into it. **(B)**

8. Once the filter is assembled and set up: Slowly pour water into the top bucket. Allow gravity to pull the water through the layers of gravel, sand, charcoal, and cloth. The filtered water will drip into the bottom bucket. This water should appear clearer and smell fresher but is not yet safe to drink. If you intend to consume the water, purify it by bringing it to a rolling boil (where the bubbles are coming from the bottom of the pot to the top) or using water purification tablets or another reliable purification method.

TIPS

- Label your buckets (especially if you're storing the filter for emergency use) clearly as "filter top" and "collection bottom."
- Let the first batch of filtered water run through and discard it—this helps remove any loose debris or dust from the new filter materials.
- Periodically check and clean the filter materials. Change them when the cotton cloth or coffee filter isn't allowing water to freely pass through.
- Replace the activated charcoal every few months to maintain filtration efficiency.
- Experiment with different filter layers. For example, adding a layer of crushed seashells can help with pH balance, while using different types of sand can affect the filtration speed and quality.

- Before pouring water into the filter, let it settle for a while. This can allow larger particles to sink to the bottom, improving the filter's efficiency.
- If the water is filtering too slowly, try adjusting the amount of filter material or the size of the hole in the top bucket. Using more coarse materials can increase flow.

CAUTIONS

- Avoid using heavily contaminated water (e.g., from industrial sources or water with visible pollutants), as this method may not remove all harmful pathogens or chemicals.
- Consider using a water testing kit to check for bacteria, parasites, and chemicals.
- To make filtered water safe for consumption, boil it to kill any remaining pathogens or use backup purification methods (like tablets).
- Understand the limitations of the filter. While it can improve water quality, it may not eliminate all contaminants, so use it as part of a larger water purification strategy, if necessary.

UNDERSTAND THE LIMITS: HOMEMADE VS. COMMERCIAL FILTERS

Misconceptions about water filtering and purifying abound, largely due to the fact that a lot of water filters on the market can filter water through pores about 0.01 micrometers in diameter, which is small enough to get rid of most dangerous waterborne pathogens. Homemade filters don't have this capacity, as it requires pressure and porous material. The best you can hope for with a homemade filtering system is to make the water look cleaner and smell better, and to remove a small percentage of some toxins prior to boiling or using it to wash with. Still, this can be a very useful system if the only available water is filled with particles and you are looking for a starting point.

16 PREPARE WATER PURIFICATION SUPPLIES

Access to clean and safe drinking water can be compromised during emergencies such as natural disasters, infrastructure failures, or contamination events. In such situations, having a reliable method for purifying water can be the difference between health and illness, or even survival and hardship.

By proactively learning about and gathering water purification supplies, individuals and families can increase their resilience in the face of unexpected challenges. This preparation not only builds confidence but also ensures that you are equipped to manage scenarios where access to clean water is limited, putting control over your own health and safety back in your hands.

Keep in mind that each person requires approximately 1 gallon per day for drinking, cooking, and basic hygiene. Be sure to stock enough supplies prepared to meet these needs over an extended period.

WHEN TO DO IT

Gather these supplies well before an emergency occurs. Preparing in advance allows you to assess your needs, acquire the right materials, and familiarize yourself with purification methods while conditions are calm. Seasonal shifts, such as the onset of hurricane season or heavy rainfall, can increase the chances of water contamination or supply disruptions. These events often cause a surge in demand, and purification supplies are among the first items to become unavailable once a disaster strikes.

MATERIALS

Having a well-rounded kit is essential for ensuring access to safe drinking water during emergencies. Here are some recommended supplies to consider including in your preparedness plan. While you don't necessarily need every item on the list, it is good to have a variety of methods available, as different situations may call for different approaches.

- **Water filters:** Portable water filters can effectively remove bacteria, protozoa, and sediment. Look for filters that have a high flow rate and can purify large volumes of water. Consider getting a few LifeStraws (small, lightweight, portable filters) as well. They don't filter large amounts of water at a time but are convenient and easily carried should you need to hit the road. **(A)**
- **Activated charcoal:** This is useful for improving taste and removing odors and some chemicals. It can be added to DIY filters (see page 123) or used in conjunction with other purification methods. **(B)**
- **Water purification tablets:** Chemical tablets, such as those containing chlorine dioxide, can kill bacteria and viruses in water. Be sure to check the expiration dates and have a sufficient supply on hand. Some tablets may contain iodine and are not good for people with thyroid issues or in long-term survival scenarios. **(C)**
- **Equipment for boiling water:** Boiling water is one of the most effective methods for killing pathogens. A portable stove or camping cook set can be used. Just make sure you have a fuel source for cooking. **(D)**
- **Water storage containers:** BPA-free plastic or food-grade water storage containers are essential for collecting and storing purified water. Ensure they are clean and have tight-fitting lids. **(E)**
- **Siphoning tools:** A siphon pump can help transfer water from one container to another, especially when filtering or collecting rainwater.
- **Water testing kit:** A testing kit can help you assess water quality for contaminants, ensuring that the purification methods you've used are effective. **(F)**

KEY ACTIONS

1. Decide whether you want to filter, boil, use purification tablets, or use a combination of all three, based on available supplies and your needs.
2. Collect water from a reliable source. If you're in an emergency, gather water from available sources like streams, lakes, or even rainwater (see page 120).

A
B
C
D
E
F

3. Purify the water using one of the following three methods:

   Filter: If using a portable water filter, read the manufacturer's instructions for assembly and use. Pour the collected water into the filter system or pump it through the system, allowing it to pass through the device into a clean storage container.

   Boil: Ensure you have a clean pot or container for boiling water and a reliable heat source (like a camping stove). Fill your pot with water and bring it to a rolling boil to kill pathogens.

   Treat with water purification tablets: Check the expiration date on the package for effectiveness. Follow the manufacturer's instructions, adding the recommended number of tablets to the collected water and letting it sit for the specified time.

4. Once the water is purified, transfer it to clean food-grade storage containers. Ensure the containers are sealed tightly to prevent contamination.

TIPS

- Watch for community alerts regarding water quality issues, such as boil-water advisories, and be prepared during these times.
- Keep purification supplies organized and easily accessible for quick use in emergencies.
- Review and refresh your supplies every six months, especially if they have been stored for a while. This will ensure you have reliable access to safe drinking water when it matters most.
- Periodically check your purification systems for clogs or wear, especially if you're using a filter. Replace filter as needed.

CAUTIONS

Drinking contaminated water can lead to a variety of symptoms, which may vary depending on the type of contaminants present. Common symptoms include:

- **Gastrointestinal issues:** Nausea, vomiting, diarrhea, and abdominal cramps are often the first signs of waterborne illnesses.
- **Fever:** A mild to moderate fever may develop as the body fights off infection.
- **Dehydration:** Symptoms like dry mouth, excessive thirst, decreased urine output, and fatigue can occur due to loss of fluids from vomiting or diarrhea.
- **Headaches:** General discomfort and headaches can arise from dehydration or the body's reaction to contaminants.
- **Muscle and joint pain:** Some contaminants can cause inflammation and pain in muscles and joints.
- **Fatigue:** General tiredness and weakness may result from the body dealing with infection or dehydration.

If any of these issues are occurring, check the quality of your water with a testing kit.

If you suspect that you've consumed contaminated water and are experiencing severe symptoms or prolonged illness, seek medical attention right away.

17 LEARN TO ACCESS HIDDEN WATER SOURCES AT HOME

In our homes, there are often overlooked sources of water that can be invaluable during emergencies when regular supplies are disrupted. For instance, a water heater can hold a significant amount of potable water, typically 30 to 50 gallons, which can be accessed by turning off the power supply and opening the drain valve at the bottom of the tank. Additionally, the toilet tank (not the bowl) can provide clean water for drinking or cooking as long as it hasn't been treated with chemical cleaners or deodorizers. Ice cubes in the freezer can be melted for drinking water. Even the water used in radiators or evaporative coolers can be used in an emergency, though they should be treated with caution and are generally unsafe for drinking without proper treatment and purification. By recognizing and safely utilizing these hidden reserves, homeowners can significantly enhance their emergency preparedness and ensure continued access to water when it's most needed.

When you combine all the sources of hidden water in your home, you could have access to 80 to 150 gallons of water. This could keep a family of four going for over a month and allow for time to establish other, more permanent water options. Just be sure to have some empty water containers ready in order to make the most of it.

WHEN TO DO IT

Accessing these alternative water sources is something you would typically do after an emergency or disaster has occurred. However, it is a good idea to identify and familiarize yourself with these potential sources in advance. Make a list of the options available in your home and learn how to access each one safely so you are prepared should the need arise.

KEY ACTIONS

WATER HEATER

1. Locate your water heater (usually in the basement, garage, or utility room).
2. Turn off the power or gas supply. For electric heaters, switch off the circuit breaker. For gas heaters, turn the gas control valve to "Off."
3. Turn off the cold water supply to prevent more water from entering the tank. (If the water supply coming in is contaminated, you don't want to let it mix with the non-contaminated water.)
4. Drain the tank: Find the drain valve at the bottom of the heater. Attach a clean hose to this valve or place a food-safe bucket underneath. Slowly open the drain valve and collect water. Be careful, as the water may be hot.
5. Filter and purify water if you suspect contamination.

TOILET TANK

1. Carefully lift the lid off the toilet tank. (Do not collect water from the bowl.)
2. Inspect the water to ensure it's clean and free of chemical cleaning tablets or deodorizers.
3. Scoop or pour the water into a sanitized food-safe container.
4. Filter and purify water if needed, especially if unsure about its cleanliness.

ICE FROM FREEZER

1. Collect ice cubes and place in a food-safe container.
2. Allow the ice to melt at room temperature or place it in a pot and gently heat it.
3. Use or store in a sanitized food-safe container.

RADIATORS AND EVAPORATIVE COOLERS

1. Ensure the heating system is turned off and cool to the touch.
2. For radiators:
    - Locate the drain valve at the bottom.
    - Place a container or attach a hose to collect the water.
    - Do not drink without treating—radiator water may contain rust, anti-freeze, or other contaminants.
3. For evaporative coolers:
    - Open the unit and locate the water reservoir.
    - Scoop or drain the water into a clean container.
    - Filter and purify before any use.

    Important: If considering using water from radiators or evaporative coolers, be cautious as this water may contain rust or other contaminants. It's typically not safe for drinking unless treated properly (see page 127). Always prioritize safer sources.

TIPS

- Label your storage containers with the date of collection and intended use.
- Store the water in a cool, dark location to prolong its freshness.

- Don't forget to check outside sources as well. Pools and ponds can provide a good water source. Depending on how they are maintained, they generally require some form of distillation to make them drinkable, rather than just purifying or filtering, but the water can be used for cleaning and sanitation.
- The average bathtub can hold approximately 60 gallons of water. This is enough to keep a family of four hydrated for about fifteen days. If you have an emergency warning in place, fill the bathtub just in case you lose power.

CAUTIONS

- When accessing water from a water heater, remember that it can be very hot. Always allow it to cool down before collecting it and handle the drain valve with care to avoid burns.
- Ensure that any containers used to collect or store water are clean and food-grade to prevent contamination. Avoid using containers that have previously held chemicals or non-food substances.

SMART WATER PLANNING WHEN SYSTEMS FAIL

When water supply systems are disrupted, the demand for accessible water sources increases, sometimes leading to competition among residents. This competition can create tension and may result in unequal access to essential resources. To mitigate the impact of such situations, it is often better to prioritize using water from home first.

By relying on your stored water supplies, you reduce the immediate demand on public or shared water sources. However, it is always good to be prepared to branch out, so make sure to keep notes and maps on hand to help you access local water sources should you need them.

18 USE A BATHTUB WATER BLADDER

A bathtub water bladder—also known as a water containment bladder—is a flexible, inflatable bag designed to fit inside a bathtub. This innovative solution allows homeowners to store water, providing an alternative water supply during emergencies or when conventional sources are unavailable. The bladder is specifically designed for bathtubs, making it a space-efficient option for water storage.

Depending on its size, a bathtub water bladder can typically hold anywhere from 30 to 100 gallons, creating a substantial emergency water source. It is generally easy to fill and drain, often featuring a simple valve for convenient access. Many bladders are made from food-grade materials, ensuring that the water stored inside remains safe for consumption and is protected against contamination.

One of the significant advantages of a bathtub water bladder is its portability; when not in use, it can be deflated and stored compactly, making it a convenient addition to your emergency preparedness plan. Overall, this product serves as a practical solution for increasing water storage capacity in your home, especially during situations where the regular water supply may be compromised.

WHEN TO DO IT

Purchase and store a bathtub water bladder as part of your emergency preparedness plan. It should be filled when emergency warnings are issued.

MATERIALS

- Water bladder (made from durable food-grade materials; look for one that fits your bathtub), with manufacturer's instructions
- Filling hose (if not already included with the bladder; food-grade)
- Drainage valve (if not already included with the bladder)
- Water treatment tablets (optional)
- Storage container

KEY ACTIONS

1. Clean your bathtub thoroughly to remove any dirt or contaminants. Rinse well to ensure no cleaning agents remain.
2. Open the bladder and inflate it according to the manufacturer's instructions. Make sure it's securely inflated. Place it flat in the tub and ensure it's properly positioned.
3. Attach a clean, food-grade filling hose to your bathtub faucet. Ensure the connection is secure and the hose is free from contaminants.
4. Begin filling the bladder with cold tap water. Monitor the water level closely and do not exceed the bladder's capacity (usually 30 to 100 gallons).
5. Once the bladder is filled to the desired level, securely close any openings or valves to prevent leaks or contamination.

6. Keep the filled bladder in your tub, and ensure to keep it cool and dark to minimize the growth of bacteria or algae. Avoid direct sunlight, which can degrade the material over time.

7. When water is needed, open the drainage valve or use the siphon pump to dispense water into a clean container for use.

TIPS

- After use, empty the bladder completely and clean it according to the manufacturer's instructions. Allow it to dry thoroughly before storing it deflated for future use.
- Periodically inspect the bladder for any signs of wear or damage. Replace it if necessary to ensure safe water storage.

CAUTIONS

- A full 60-gallon bladder can weigh up to 500 pounds—ensure the tub is structurally sound, that your bathtub and the surrounding area can support this weight, and that it is located in an accessible area. It will be extremely hard to move once full.
- Use a bladder made from food-grade materials to avoid contamination. Check for any certifications or labels that indicate the bladder is safe for storing drinking water.
- If you plan to store water for a long time, consider treating it with water purification tablets or regularly changing the water to prevent bacterial growth or algae.
- Inspect the bladder regularly for any signs of wear, damage, or leaks. A compromised bladder could lead to water damage in your home.
- Store the filled bladder in a cool, dark place, as prolonged exposure to sunlight and heat can degrade the bladder material and promote bacterial growth.

19 BUILD A SOLAR STILL

A solar still is a device that uses sunlight to purify water through evaporation and condensation. It's typically used to purify water by effectively removing contaminants, salts, and impurities that other purification methods won't. This makes it particularly useful in regions with brackish or saline water.

Solar stills can be designed and constructed using local materials, allowing for customization based on available resources. This flexibility enhances their practicality in various environments and emergency scenarios.

On average, a good solar still can produce anywhere from 1/4 cup to 1 cup of potable water per hour. This may be enough to help you survive, but it's not enough to rely on, so it's important to use it in conjunction with other water-sourcing methods. Having multiple solar stills operating at once is also a good option if you have the materials available.

WHEN TO DO IT

Build a solar still after an emergency occurs, but gather the materials in advance and store them in an accessible location. It's also a good idea to practice using one before you actually need it.

MATERIALS

- Shallow (2 x 2 feet, approx.) basin or container (plastic, glass, or metal)
- Smaller food-grade container or cup
- Clear plastic sheet or glass pane large enough to extend beyond the edges of the shallow container
- Small weight or stone

KEY ACTIONS

1. Choose a location that receives direct sunlight for most of the day. (The more sun exposure, the more drinking water you will filter.)
2. Place the shallow container on a flat, level surface.
3. Pour the contaminated or brackish water into the shallow container. Add enough water for evaporation but avoid overfilling.
4. Position the smaller container or cup in the center of the shallow container. Ensure it is stable and not floating. This will collect the distilled water as it condenses.
5. Cover the container with the clear plastic sheet or glass pane. The cover should extend beyond the container edges to trap heat effectively.
6. If using a plastic sheet, place a small weight, such as a rock, in the center of the cover, directly above the inner container. This creates a low point for the condensed water droplets to gather under the cover and drip into the collection container.
7. If necessary, secure the edges of the cover with soil or stones to prevent heat loss and keep the setup stable.
8. Leave the solar still in place for several hours. The heat will cause the water to evaporate, and as it cools on the underside of the cover, it will condense and drip into the collection container.
9. After several hours, check the small container for distilled water. Carefully remove the cover to extract the clean water, and then reseal to collect more distilled water.

10. Repeat as needed, adding more contaminated water to the shallow container or refreshing the setup.

TIPS

- Use a wide, shallow container with a larger surface area, as this allows more water to evaporate quickly.
- Ensure the cover fits well and seals tightly around the edges to minimize heat loss and maximize efficiency.
- If you can, paint the bottom of the shallow container black or use a dark-colored container. This will absorb more heat from the sun and speed evaporation.
- If it's windy, try to shield the still from direct wind to prevent heat loss and ensure better efficiency.
- Be patient. Distillation can take time, especially in cooler weather or less sunny conditions.

CAUTIONS

- Avoid using highly contaminated or toxic water sources. Solar stills are effective for removing salts and some impurities, but they may not eliminate all harmful chemicals, pathogens, or heavy metals.
- Do not rely solely on solar stills. In emergency situations, have other water purification methods available as backups (see page 127).

20 LOCATE WATER SOURCES

Urban areas are usually intricately connected to centralized water supply systems, and any disruption—whether due to natural disasters, infrastructure failures, or maintenance—can significantly impact access to clean water. In such situations, having an understanding of alternative water sources, such as rivers, lakes, or public fountains, becomes invaluable.

WHEN TO DO IT

Identify nearby water sources well before any emergencies arise. That way you'll have time to thoroughly gather information without the pressure and urgency that come with a disaster scenario. Be sure to anticipate seasonal changes that may affect water availability, such as droughts or periods of heavy rainfall that could lead to flooding.

MATERIALS

- Local map (topographic maps are good but GPS apps work just as well)
- Notebook and pen

KEY ACTIONS

1. Research local geography and water infrastructure. City maps, online resources, and government websites can provide information about natural and manufactured water sources. Many apps offer features that highlight water sources and can provide directions to them. Look for rivers, lakes, ponds, reservoirs, and public water fountains.
2. Visit nearby parks and recreational areas, as they often contain ponds, lakes, or streams. These areas are typically easy to access and may offer additional resources like restrooms and public fountains.

3. Reach out to local government offices or water departments. They can provide detailed information about municipal water sources, including their safety and accessibility.
4. Engage with local community groups, neighborhood associations, or social media platforms. People who have lived in the area for a long time can offer valuable insights and may know of lesser-known water sources.
5. Once you've identified potential water sources, verify their quality. For natural sources, check for signs of contamination or pollution. Local environmental agencies can provide water quality reports for public water bodies.
6. Create a personal or community map documenting these water sources. Include details like location, source type, accessibility, and any relevant notes about water quality or restrictions.

TIPS

- Periodically review and update your map to account for changes in the urban landscape, water availability, or quality.
- Note the locations of stores, markets, or facilities that sell bottled water or have water refill stations. Not all water sources need to be natural.

CAUTIONS

- Take the time to purify water from external sources: 80 percent of water found above the ground is contaminated.
- Some water sources may be protected or restricted by local laws or regulations. Get permission to access and use the water from private properties or protected reserves.
- Consider the safety and accessibility of the water source. Avoid areas that require trespassing or pose physical dangers, such as steep banks, strong currents, or unstable ground
- Prioritize safety over access to water. Follow guidelines and instructions from authorities regarding safe water sources and distribution points.

21 SET UP A HANDWASHING STATION (TIPPY TAP)

A handwashing station plays a crucial role in maintaining hygiene and preventing the spread of diseases during crises. One of the most effective and resource-efficient solutions for an emergency handwashing station is the "tippy tap"—a simple, ingenious device designed to promote hand hygiene while conserving water. It typically consists of a container with a small hole, suspended by a rope, that allows the container to tip and pour water when a user presses a foot pedal or lever. This hands-free operation minimizes the risk of contamination, as users do not need to touch the tap with their hands.

It can be built with readily available materials such as plastic bottles, string, and sticks, making it practical and easy to implement in various environments. Its low-cost design makes it an affordable option where resources are limited.

WHEN TO DO IT

Make a handwashing station, such as a tippy tap, during the planning and preparation phase of your emergency preparedness plan. This ensures you're ready well before any actual emergency occurs. Practice assembling and using it during emergency drills to ensure it's operational and familiar to everyone involved.

MATERIALS

- Clean plastic bottle or jerry can (around 5 liters, with a cap)
- String or rope
- 2 sturdy sticks or poles
- Small stick or piece of wood
- Nail or sharp object
- Bar of soap or liquid soap
- Nails/screws
- Hammer or screwdriver

KEY ACTIONS

1. Use a nail or other sharp object to make a small hole near the top of a clean plastic bottle or jerry can, just below the cap. This will allow the water to flow out. **(A)**
2. Fill the bottle with water to test the hole size. The hole should allow a slow, steady stream. Adjust the hole size, if needed. **(B)**
3. Insert two sturdy sticks or poles into the ground about 1 meter apart. Make sure they are firmly planted and upright. This is the main support structure. **(C)**
4. Suspend the container horizontally between the two sticks using string or rope. Tie the rope through the bottle's handle or around its body so it can easily tilt forward and return to its original position when released. **(D)**
5. Tie a bar of soap to one of the sticks for easy access. **(E)**
6. Create a foot pedal mechanism by tying a string to the neck of the bottle and attaching the other end to a small stick on the ground, which will act as the foot pedal. Stepping on the stick should tip the container forward. **(F)**
7. Fill the container with clean water. Make sure the cap is closed tightly so water exits through the small hole only when the bottle tips.
8. To use:
   - Step on the foot pedal to tilt the container and let water flow.
   - Lather your hands with soap and rinse in the flowing water.
   - Release the pedal to stop the water flow. **(G)**

TIPS

- Place the tippy tap near the entrance of your home or shelter, ensuring it's easily accessible to everyone.

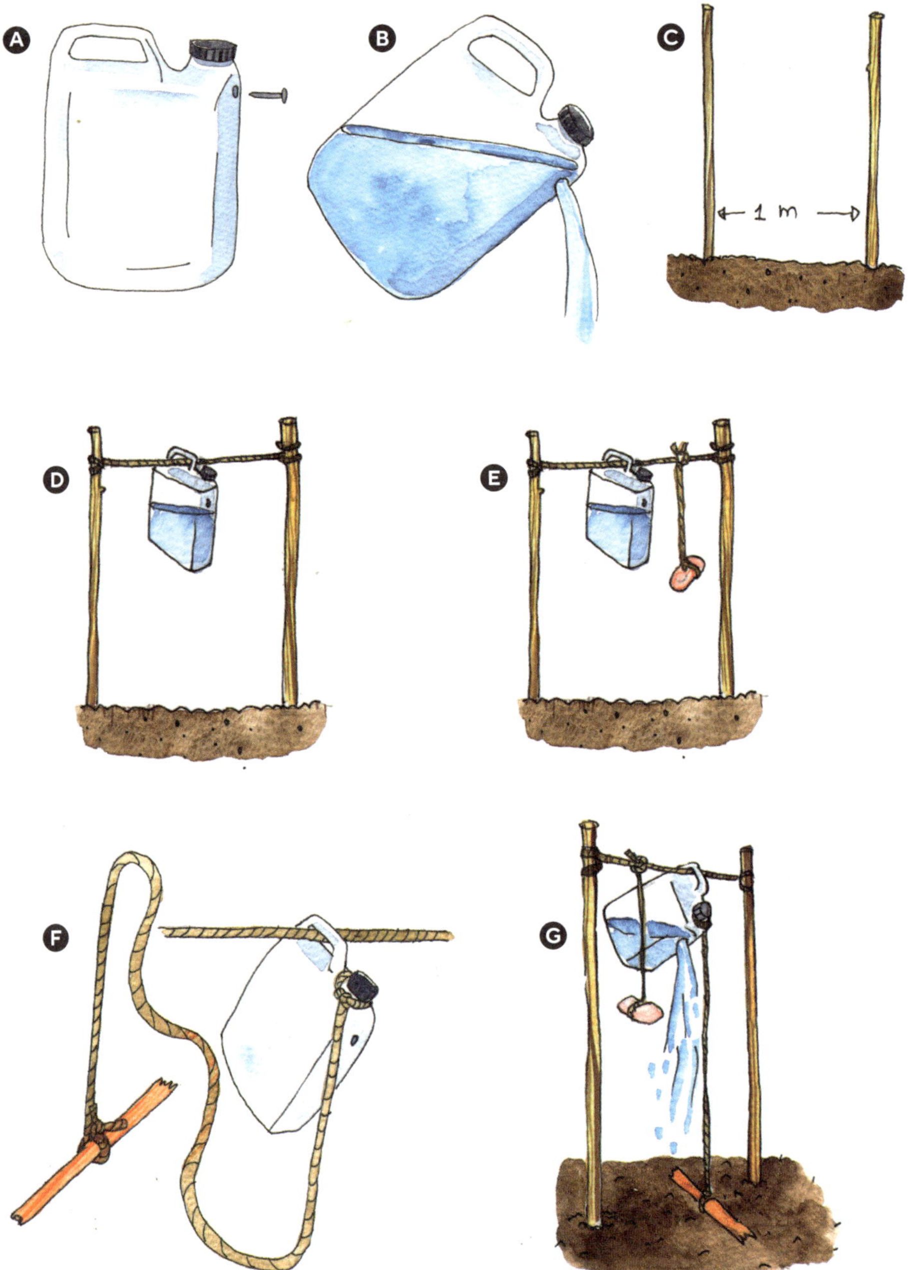
A
B
C
1 m
D
E
F
G

- Make sure the string or rope holding the container is strong enough to handle repeated use and the weight of the filled container.
- Use the foot pedal gently to control the flow of water and avoid wastage.
- Make sure everyone, especially children, knows how to use the tippy tap properly to maintain hygiene.
- Keep an eye on the water level and refill the container as needed to ensure a continuous supply.

CAUTIONS

- Ensure the water used in the tippy tap is clean and safe for handwashing to prevent contamination and potential health risks.
- Regularly check the stability of the frame and the security of the ropes or strings to avoid accidents or the structure collapsing.
- Clean the container, soap holder, and surrounding area regularly to maintain hygiene and prevent the buildup of dirt or bacteria.

22 RECYCLE GREYWATER FOR TOILETS OR GARDEN

During emergencies, access to fresh water can become limited. Recycling greywater—water from sinks, showers, and laundry—can help conserve this critical resource. While not safe for drinking, greywater can be used for non-potable purposes, such as flushing toilets or watering gardens, significantly reducing your dependence on fresh water.

In fact, greywater can contain nutrients that are beneficial to plants, making it especially useful for maintaining food-producing gardens during a crisis.

There are several approaches to greywater recycling. Some involve complex plumbing systems that require professional installation but offer long-term benefits, including reduced utility costs. Others are simple DIY solutions that target specific sources of greywater. While these may be less efficient, they can still be highly effective in emergency situations. Whatever option you choose, know that it is aiding in your self-sufficiency.

In this project, I'll outline how to install a mid-level greywater recycling system. I'll also briefly cover simpler options for those who need a faster or more temporary solution. The moderate greywater system may seem daunting, but if you make a thorough plan and walk through it step by step, there is no reason why you can't easily install this handy system on your own. If in doubt, consult a professional plumber or chat with the person behind the counter at your local hardware or plumbing store. This system will not only save you money but will also give you the peace of mind that you have systems in place to utilize most of your water waste if water becomes scarce.

WHEN TO DO IT

Incorporate a greywater recycling system into your emergency preparedness plan well before an emergency occurs.

MATERIALS

- **Diverter valve:** A diverter valve is a plumbing device that redirects the flow of water from one outlet to another. In the context of greywater recycling, a diverter valve is used to switch the flow from down the drain to instead being directed into a recycling system. This allows you to control when and where the greywater is used, such as directing it to a storage tank for later use in toilets or gardens. You can find diverter valves at most hardware stores, plumbing supply shops, or online retailers. They're available in various sizes and types, so it's important to choose one that matches your existing plumbing setup and the specific requirements of your greywater system. If you're unsure which type to get, a conversation with a plumbing professional can help you select the best option.
- **Pipes and fittings:** The type of fittings you need will depend on your specific plumbing setup and the materials you're working with. Selecting the right fittings involves matching them to the type and size of the pipes you're using as well as ensuring compatibility with any existing plumbing. It's important to have a clear plan to ensure you have all the necessary components. Some commonly used fittings are:
  - **PVC or ABS fittings:** These are often used with plastic pipes and include elbows, tees, couplings, and reducers. They're durable, resistant to corrosion, and easy to work with.
  - **Compression fittings:** Useful for connecting pipes without the need for soldering or gluing. They create a tight seal and are often used for connections that may need to be disassembled.
  - **Barbed fittings:** Ideal for flexible hoses. These fittings have barbs that grip the inside of the hose, often secured with a clamp.
  - **Threaded fittings:** These include adapters, unions, and caps with threaded ends, allowing for easy assembly and disassembly.
  - **Hose clamps:** These are often used to secure flexible hoses to barbed fittings, ensuring a leak-proof connection.

- **Check valves**: These fittings prevent backflow, ensuring that greywater doesn't flow back into the freshwater system.

- **Filters**
- **Storage tank**
- **Pump (optional):** If gravity flow is insufficient for your setup, a pump may be needed to move the water to its destination.
- **Overflow mechanism**
- **Backflow-prevention device**

TOOLS

- Wrenches
- Plumber's tape
- Tape measure

KEY ACTIONS

1. Make a plan. Determine which sources of greywater (e.g., showers, sinks) you will use and decide where the recycled water will be applied (e.g., toilets, gardens).
2. Design the system layout. Map out the flow of greywater, including where the diverter valve, piping, and storage tank will be positioned.
3. Install the diverter valve at the appropriate location in your plumbing to redirect greywater from the drain to the recycling system. Some common locations are:
    - Under the kitchen sink
    - Under the bathroom sink
    - Shower drain

- Washing machine outlet pipe
- Bathtub drain

4. Install a backflow-prevention mechanism. Some typical locations include:
    - At the diverter valve
    - Just after each water source
    - Before the storage tank
    - At the irrigation connection

5. Connect pipes from the greywater sources to the storage tank, ensuring all connections are secure and leak-free.
6. Add filters before the greywater enters the storage tank to remove debris and particles.
7. Position the storage tank in a location that allows for easy access and integration with the rest of the system.
8. Install an overflow mechanism on the tank to divert the excess water to a safe place to avoid flooding.
9. Attach pipes or hoses from the storage tank to the end-use applications, such as garden irrigation lines or toilet connections.
10. Run water through the system to check for leaks and ensure everything is functioning as intended.

SIMPLE GREYWATER RECYCLING OPTIONS

The following methods vary in ease and efficiency but can be a good starting point if you're looking to experiment with greywater recycling without committing to a fully installed system.

Greywater buckets: The simplest method involves manually collecting greywater in buckets from sinks or showers and then using it to water plants or flush toilets. This requires no installation but does involve manual effort.

Shower diverter tubes: For showers, you can use a diverter tube that connects to the shower drain, directing water to a nearby garden or storage container. It's easy to set up and doesn't require major plumbing changes.

Laundry-to-landscape system: This method involves using a simple diverter valve on your washing machine's drain hose. It can redirect greywater directly to your garden. The setup is relatively straightforward, often requiring just a few hoses and a diverter valve.

Portable greywater systems: Some commercial products are designed to be easy to install and use. These systems typically involve a small pump and filter, allowing you to divert and use greywater with minimal plumbing work.

TIPS

- If you're using greywater for irrigation, ensure that the soaps and detergents in your household are biodegradable and safe for plants. Regularly check the health of your plants to ensure they're responding well to the water.
- Clearly label the diverter valve and associated piping to avoid confusion and misuse, especially if multiple people are using the system.
- Start small. If you're new to greywater systems, consider a simple setup, like a laundry-to-landscape system, to get accustomed to the process, and make adjustments as needed.

CAUTIONS

- Before setting up a diverter valve, familiarize yourself with local plumbing codes and regulations regarding greywater use. Compliance is essential to avoid potential fines or issues.
- Use greywater promptly to prevent it from becoming stagnant and developing odors. If stored, ensure the container is sealed and not left for extended periods.
- Clean filters and check connections regularly to ensure the system operates smoothly. This prevents clogs and maintains water quality.

PART 3

FOOD

IN AN INCREASINGLY UNCERTAIN WORLD, the importance of good food growth and storage projects in emergency preparedness cannot be overstated. In a survival situation, we can last approximately twenty-one days without food, but our dependence on three meals a day is a comforting habit and keeps us in peak condition both mentally and physically. Whether facing natural disasters, economic downturns, or unexpected disruptions, knowing how to grow and store your own food can provide a vital safety net if an emergency situation becomes prolonged and stretches from days into weeks.

Cultivating your own food empowers you to take control of your and your family's nutrition, ensuring access to fresh, wholesome ingredients even in times of crisis. A well-planned garden, whether it's in your backyard, on a balcony, or even indoors, can serve as a sustainable food source that buffers against potential shortages.

Once your plants yield a bountiful harvest, effective storage techniques become essential. Learning how to can, freeze, or dry your produce means that you can preserve your hard-earned food, extending its shelf life and ensuring that it remains accessible during emergencies. A well-stocked pantry filled with preserved homegrown goods offers peace of mind, allowing you and your family to weather any storm with confidence.

In this section, we'll delve into practical strategies for growing and storing food that will enhance your emergency preparedness. By equipping yourself with these skills, you can create a resilient foundation, ensuring that you are ready to meet challenges head-on, no matter what life throws your way.

PROJECTS

23 BUILD A ROCKET STOVE FOR COOKING

A rocket stove is a highly efficient, wood-burning cooking stove designed to use small amounts of fuel to produce a hot, clean-burning flame. It's especially useful for emergency preparedness because it's simple to build, fuel-efficient, and doesn't require electricity or gas.

There are several ways to build a rocket stove, but all follow the same basic principles. Every rocket stoves consists of three main parts: a combustion chamber, a chimney, and a cooking surface. The design allows small amounts of fuel, such as twigs or small branches, to be fed into the combustion chamber, where they are ignited. The chimney creates a strong upward draft that enhances airflow and accelerates the combustion process, resulting in a hotter fire with less smoke. This efficient airflow also ensures that the heat is directed toward the cooking surface, allowing food to cook quickly with very little fuel.

In this section, I include two designs. I recommend building the brick version as part of your emergency preparedness setup. However, it's also a good idea to practice the metal-can design so you are prepared to build one quickly if you are ever away from home in a survival situation.

WHEN TO DO IT

Construct the rocket stove at any time. It isn't complicated and can be done with materials readily available around the home.

BRICK ROCKET STOVE

MATERIALS

- 30 house bricks

KEY ACTIONS

1. Construct a small, two-foot-high chimney using bricks. Leave a gap at the bottom on one side to serve as the fuel intake. The chimney should be only one brick wide.
2. Start the fire: Place small twigs and sticks into the gap at the base and light the fire inside the chimney. As the fire burns, the heat and smoke will funnel upward, creating a strong draft.
3. Place a metal grid or grate on top of the chimney to support pots or pans. This is optional but helps balance cookware more securely.
4. As you cook, continue to feed small, dry twigs or sticks into the fuel gap as needed to keep the fire going. Avoid overloading the chimney, and use consistent-sized fuel for steady heat. Monitor the flame while cooking to ensure even heat and prevent the fire from going out.

METAL ROCKET STOVE

MATERIALS

- 2 tin cans (1 large coffee can and 1 smaller soup can)
- Metal piping (3 inches in diameter) or a third smaller tin can (standard soup can) to serve as a "feeding tube"
- Insulating material (ash, dirt, or small rocks)

TOOLS

- Metal snips or a sharp knife
- Thick work gloves

KEY ACTIONS

1. Cut a hole in the sides of both cans to serve as the feed tube. It needs to be big enough to hold the third smaller can or metal piping snugly. **(A)**
2. Cut four 1-inch squares in the top rim of the larger can to rest your pot or pan on. **(B)**
3. Place the smaller can inside the larger can and line up the holes. **(C)**
4. Insert the feeding tube into the two holes.
5. Fill the gap between the two cans with the insulating material to retain heat. **(D)**

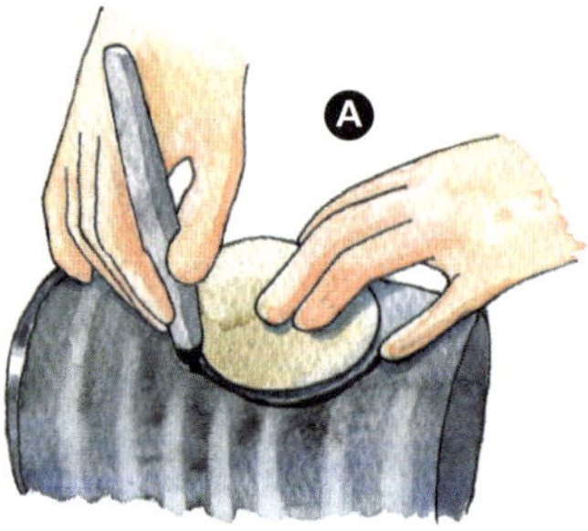

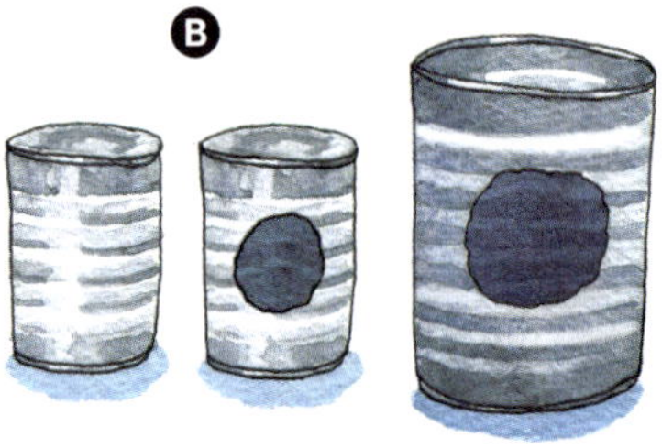

C

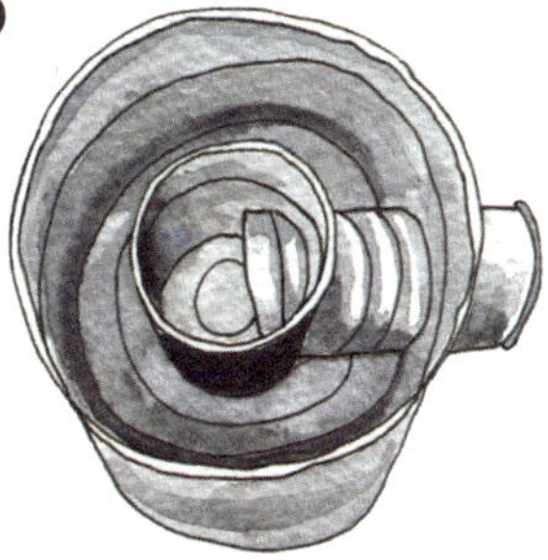

D

E

6. Once assembled, the stove can be used by adding fuel, such as small sticks, through the feed tube, igniting it, and allowing the chimney effect to enhance combustion and heat output. **(E)**

TIP

- Use small pieces of dry wood as fuel to maximize efficiency and reduce smoke.

CAUTIONS

- Wear thick work gloves when using metal snips and assembling the feed tube to prevent cuts.
- Set up the stove on a stable, nonflammable surface.
- Always use your rocket stove outdoors in a well-ventilated area to ensure safety.
- Handle the stove and cooking surfaces carefully, as they can get very hot.
- Keep flammable materials away from the stove to minimize fire risks.
- Monitor the stove during use to prevent overheating.
- Ensure the stove is safely extinguished after cooking.
- Be aware of local regulations regarding outdoor cooking and fire safety.

24 STOCK AN EMERGENCY PANTRY

An emergency pantry is a key part of preparedness, providing essential food and supplies during disruptions. A well-stocked pantry helps reduce stress in a crisis and allows you to stay self-sufficient if stores or aid are unavailable. Beyond just food, it offers peace of mind and a sense of stability in uncertain times.

To build one, think strategically about your family's needs. Start with nonperishable staples like canned vegetables, beans, and grains. Add items like dehydrated fruits, nut butters, and powdered milk for nutrition and comfort.

The key is to choose high-quality, long-lasting foods that will sustain you both physically and mentally—and don't forget to include some family favorites. Remember: It's not just about survival; it's about maintaining morale through adversity.

WHEN TO DO IT

Rather than building your emergency pantry all at once, which can be expensive, put away a little extra every time you go shopping. If you start now, you'll soon have enough stores to sustain you and your family over an extended period of time.

MATERIALS

- **Canned goods:** Canned goods offer convenience and essential nutrients. Canned vegetables are an easy way to add vitamins and minerals to your meals and can be used in soups or served as side dishes. Canned beans are rich in protein and require no cooking (but can be more flavorful if heated), while canned meats like tuna or chicken provide high-quality protein to help sustain energy. Additionally, canned soups and broths can serve as a base for comforting, nourishing meals.

- **Dried goods:** Dried goods have a long shelf life and offer great versatility. Rice is a filling carbohydrate that pairs well with a variety of proteins and vegetables, while pasta is a quick, adaptable base for countless meals. Oats are rich in fiber and energy, and can be prepared in multiple ways. Dried fruits provide a natural source of sweetness, making them ideal for snacking or enhancing meals.
- **Freeze-dried and dehydrated foods:** Freeze-dried and dehydrated foods are excellent additions to an emergency pantry thanks to their lightweight nature and long shelf life. Freeze-dried fruits retain their flavor and nutrients, making them perfect for snacking or adding to cereals and smoothies. Dehydrated vegetables are a convenient way to enhance soups, stews, and casseroles, as they rehydrate easily when needed. Freeze-dried meals also offer a variety of ready-to-eat options that require only water, providing hearty, nourishing meals with minimal effort.
- **High-calorie foods:** High-calorie foods are essential for an emergency pantry because they provide quick energy and lasting sustenance. Nut butters are packed with healthy fats and protein. Granola bars, protein bars, and trail mix offer portable, calorie-dense options that combine nuts, dried fruits, and grains for a balanced energy boost.
- **Luxury foods or treats:** Luxury foods, such as sweeteners and baking goods, can elevate your emergency pantry by providing comfort and variety during stressful times. Honey, a natural sweetener with an indefinite shelf life, is perfect for enhancing meals, beverages, or baked goods. Flour and baking mixes for pancakes, muffins, or bread make it easy to prepare comforting foods. Including these items not only adds flavor but also brings a sense of indulgence and normalcy when it's needed most.

Being the survivalist I am, I can't overlook the essentials that go hand-in-hand with your food supply. A reliable way to prepare meals is essential: a portable stove, along with an ignition source and fuel, is a must. Beyond food, be sure to include water, first-aid supplies, flashlights, and extra batteries in your pantry kit. This well-rounded approach ensures you're not just ready to eat but fully equipped to handle whatever comes your way.

KEY ACTIONS

1. Assess your needs. Consider how many people you are preparing for and their dietary preferences or restrictions. Decide how long you want the pantry to sustain you, typically aiming for at least three days to two weeks, depending on your comfort level.
2. Make a food list. Consider your family's preferences. Involve them in the process to ensure you are stocking foods that will get eaten.
3. Find a storage space. Choose a cool, dry, and dark area in your home, such as a basement, closet, or existing pantry. Make sure it's easily accessible and suitable for organization.
4. Obtain containers and shelving. Use airtight containers, bins, and shelves to protect food from pests and keep it organized.
5. Start stocking up. Purchase nonperishable items in bulk to save money and build a solid supply.
6. Use labels. Clearly mark expiration dates and contents on containers for easy identification.
7. Plan meals. Research and write down easy recipes using your pantry items to make meal prep easier during emergencies.

TIPS

- Diversify your stock. Include a variety of food types to ensure balanced nutrition. Stock up on grains, proteins (like canned meats and legumes), fruits, vegetables, and snacks. This diversity not only provides essential nutrients but also keeps meals interesting.
- Prioritize long shelf life. Focus on foods with staying power, such as canned goods, dried beans, rice, pasta, and freeze-dried items. Regularly check expiration dates and rotate your stock to use older items first.

- Create an inventory system. Maintain a written or digital inventory of your pantry items, including quantities and expiration dates. This helps you keep track of what you have so you can avoid overbuying and ensure you use items before they expire.
- Have a rotation system that allows for food that might be about to expire to be used and replaced.

CAUTIONS

- It may be tempting to dip into the emergency pantry when you are short on time or groceries, but it's important to maintain the mindset that this food is reserved for emergencies only. If you do take food out, be sure to replace it at the first chance you get.
- Store food in a cool, dark, and dry place to prevent spoilage and maintain quality. Avoid areas with fluctuating temperatures or high humidity.
- Consider dietary restrictions and allergies of all family members when stocking your pantry. Avoid items that could cause adverse reactions.
- Keep the pantry area clean to prevent pests and contamination.
- Don't overcrowd your storage space, as this can make it difficult to access items and lead to spoilage or waste.

EMERGENCY PANTRY ESSENTIALS: A CHECKLIST

- ☐ Beans and legumes, dried and canned
- ☐ Chocolate
- ☐ Coconut milk, canned
- ☐ Coffee and tea
- ☐ Cooking oils
- ☐ Fish, canned
- ☐ Flour
- ☐ Fruits, canned
- ☐ Fruits, dried
- ☐ Granola bars
- ☐ Honey
- ☐ Oats
- ☐ Pasta
- ☐ Peanut or other nut butters
- ☐ Powdered milk
- ☐ Rice
- ☐ Salt
- ☐ Soups and stews, canned
- ☐ Spices
- ☐ Sugar
- ☐ Vegetables, canned

25 LEARN HOME CANNING TECHNIQUES

Canning can be a vital component of food storage in emergency preparedness, as it allows you to preserve fresh fruits, vegetables, and other perishable items for extended periods, often up to a year or more, without the need for refrigeration. This ensures a reliable source of nutritious food during emergencies, when access to fresh produce may be limited or unavailable. By canning seasonal harvests, you can stockpile essential nutrients, helping to maintain health and well-being during crises.

Additionally, canning offers the flexibility to store a variety of foods, including meats, soups, sauces, and salsas, which can provide diverse meal options in times of need. This variety not only helps prevent food fatigue but also ensures that families can enjoy familiar tastes even in challenging situations. To ensure safety and quality, always follow tested recipes and adhere to USDA guidelines.

WHEN TO DO IT

Can during peak harvest seasons, typically in late summer and early fall, when fruits and vegetables are at their freshest and most flavorful. It's also wise to can food when you have an abundance of produce, both to prevent waste and to ensure a well-stocked pantry for emergencies.

MATERIALS

- Canning (Mason) jars
- Canning lids and bands
- Water bath canner and/or pressure canner
- Jar lifter
- Canning funnel
- Clean cloth or paper towels
- Large pot for boiling water
- Food to be canned

KEY ACTIONS

1. Wash and prepare food according to the specifications in your recipe. This may include peeling, chopping, blanching, or cooking vegetables and fruits. Follow your recipe carefully when mixing in ingredients like sugar, vinegar, spices, flavorings, or pectin.
2. Place jars in a boiling water bath for ten minutes to sterilize them. Alternatively, you can run them through a dishwasher cycle. Keep them warm until you're ready to fill. **(A)**
3. Soak the canning lids in hot (not boiling) water for about five minutes. This softens the sealing compound to help ensure a good seal. Note: Some newer lids do not require preheating—check the manufacturer's instructions. **(B)**
4. Using a canning funnel, fill the sterilized jars with the hot food, leaving the appropriate headspace as specified in your recipe (usually 1/2 inch to 1 inch). **(C)**
5. Release any trapped air bubbles by gently sliding a nonmetallic spatula around the inside of the jar. **(D)**
6. Use a clean cloth or paper towel to wipe the rims of the jars clean to ensure they are free of food residue, which can prevent a proper seal. **(E)**
7. Place the prepared lids on the jars and screw on the metal bands just until they are fingertip-tight (do not overtighten). **(F)**
8. Process the jars according to recipe specifications. **(G)**
    - For water bath canning (high-acid foods like tomatoes and fruit): Place jars in the canner, ensuring they are covered by at least 1 inch of water. Bring to a rolling boil and process for the full time specified in your recipe.
    - For pressure canning (low-acid foods like meats, beans, and most vegetables): Follow your canner's manufacturer's instructions for correct pressure levels and processing times based on your altitude and the type of food being canned.

9. Once processing is complete, turn off the heat. For pressure canners, allow the canner to depressurize naturally.

10. Carefully remove the jars using the jar lifter and place upright on a clean towel or cooling rack. Let them cool undisturbed for twelve to twenty-four hours. **(H)**

11. After the jars have cooled, check the seals by gently pressing down on the center of each lid. If it doesn't flex or pop back, the jar is sealed properly. If the lid pops up and down, the jar did not seal—refrigerate the jar and consume its contents within a few days. **(I)**

12. Label the jars with the contents and the date. Store them in a cool, dark, and dry place (ideally 50°F to 70°F). **(J)**

TIPS

- Avoid overfilling jars, as this can prevent a proper seal and lead to food spoilage.
- Label your jars with the contents and date of canning.
- Store jars in a cool, dark place to maintain quality.
- Properly canned food can last for a year or more, but it's best to consume them within a year for optimal taste and texture.

CAUTIONS

- Never can dairy products, flour-thickened sauces, or other items not recommended for home canning.
- Always use tested, approved canning recipes from reliable sources such as the USDA or the National Center for Home Food Preservation. These recipes provide the correct processing times and methods to prevent foodborne illness.
- Always follow guidelines for acidification (adding lemon juice or vinegar) to ensure that the pH level is safe for canning.

A
B
C
D
E
F
G
H
I
J
Date

26 STORE A NINETY-DAY EMERGENCY MEAL PLAN WITH A ROTATION SCHEDULE

Although most off-grid emergency scenarios tend to be rectified within twenty-four hours to three days, some disasters can stretch into weeks and months. Creating a ninety-day emergency meal plan with a rotation schedule will ensure that you and your family are prepared for unforeseen circumstances while maintaining a balanced diet.

By establishing a rotation schedule, you can effectively manage your inventory, preventing items from expiring and reducing food waste. This approach also helps you create a sense of normalcy and routine during stressful times, as having a structured meal plan can alleviate the anxiety of meal preparation. Moreover, a well-thought-out plan enables you to incorporate comfort foods and essentials, making it easier to maintain both physical health and emotional well-being when faced with challenges.

By following these steps, you can successfully create a comprehensive ninety-day emergency meal plan that will keep you and your family well fed if your disaster scenario stretches from days into months.

WHEN TO DO IT

Create a ninety-day emergency meal plan during the initial stages of your overall preparedness plan.

MATERIALS

- Notebooks or digital tools
- Meal-planning templates
- Storage containers
- Labels
- Kitchen scale and measuring tools
- Food items

KEY ACTIONS

1. Assess your needs. Discuss dietary preferences, allergies, and any specific nutritional needs with your family.
2. Gather materials. Collect notebooks or digital tools for documentation. Find or prepare an inventory list template. Acquire meal-planning templates. Gather canning and preservation supplies, if applicable. Obtain storage containers and labels.
3. Inventory current supplies. Check your pantry and list all nonperishable items you may wish to use for the project. Organize items into categories (grains, proteins, fruits, vegetables, snacks) to identify gaps.
4. Plan your meals. Outline a ninety-day meal plan template, including spaces for breakfast, lunch, dinner, and snacks. Choose recipes that utilize the foods you have on hand and those you plan to purchase. Aim for variety and balanced nutrition as well as including some favorites of each family member. Ensure that the plan includes treats or snacks to help maintain morale.
5. Make a shopping list. Based on your meal plan and inventory, create a shopping list of items needed to complete your ninety-day supply. Make sure you focus on nonperishable items that have a long shelf life.
6. Source supplies. Shop for the items on your list, considering bulk purchases for savings. Ensure that the items you purchase have sufficient shelf life before their expiration. If possible, can, freeze-dry, or dehydrate your own food supplies, as it will save you money.
7. Organize your pantry so that similar items are grouped together for easy access. Store bulk food in airtight containers and label with contents and expiration dates. Keep an updated inventory of your food supplies.

8. Schedule weekly checks of perishable items to assess freshness and identify items nearing their expiration dates. Designate a system for using older items or items that may spoil faster than others first. Divide items into three categories:
   - Highly perishable: foods that spoil quickly
   - Moderately perishable: items with a slightly longer shelf life
   - Less perishable: items with a year or longer shelf life
9. Practice "First In, First Out" (FIFO): Organize your fridge and pantry so that older items are at the front and used first, reducing the likelihood of spoilage.
10. Use visual aids. Label items with expiration dates to keep track of freshness. Use transparent containers to easily see what needs to be used soon.

TIPS

- Try preparing some meals from your plan to ensure familiarity with recipes and cooking methods.
- Stay informed about new recipes, preservation methods, and food safety guidelines to keep your emergency meal plan relevant and effective.
- Properly store food in a cool, dry, and dark place to maximize shelf life.

CAUTIONS

- Ensure that your meal plan includes a variety of food groups to meet nutritional needs. Relying too heavily on one type of food can lead to deficiencies over time.
- Be vigilant about the expiration dates of the items you stock. Regularly check and rotate your inventory to prevent spoilage and waste.
- Consider any food allergies or sensitivities when selecting items for your meal plan. This will help avoid health issues during emergencies.

27 START A BACKYARD SURVIVAL GARDEN

With rising food prices and potential supply chain disruptions, growing your own food allows you to take control of your nourishment and reduce dependence on grocery stores. A survival garden provides a reliable source of fresh produce, which can be crucial during emergencies or times of scarcity.

The size of your backyard garden will depend on the space available to you. Some herbs, vegetables, and fruits can be grown in an area as small as an apartment balcony, but to maximize your gains, I recommend an area no smaller than 200 square feet for a small family.

Once your garden starts producing, be prepared to preserve excess food through canning (see page 168), freezing, or drying. This extends the availability of your harvest during lean times. I like to share the abundance throughout my neighborhood, as it builds a sense of community and encourages a bartering-type system.

Ultimately, a backyard survival garden is a practical solution for enhancing self-sufficiency, saving money, and ensuring access to fresh, healthy food, making it an important addition to any household.

WHEN TO DO IT

Establish a backyard survival garden well in advance of any potential crisis. Begin planning your garden at least a few months before you anticipate needing it. This allows time for research, soil preparation, and selecting the right crops for your climate. Keep in mind your local growing seasons. Depending on your region, certain vegetables and herbs thrive in spring, summer, or fall. Planting according to these seasons will ensure a bountiful harvest.

MATERIALS

- Seeds or seedlings (a variety of vegetables, herbs, and fruits suitable for your climate and growing season)
- Quality garden soil or a mix of topsoil, compost, and organic matter
- Compost and/or fertilizer
- Watering equipment (hoses, watering cans, or drip irrigation system)
- Mulch (straw, wood chips, or grass clippings)
- Fencing (optional)
- Plant supports (trellises, stakes, or cages)
- Pest control (neem oil, insecticidal soap, or natural predators like ladybugs)
- Plant labels
- Gardening gloves

TOOLS

- Shovel
- Rake
- Hoe
- Trowel
- Pruning shears

KEY ACTIONS

1. Plan your garden. Begin by assessing how much space you have available. Consider factors like sunlight exposure, shade patterns, wind direction, water access, and overall accessibility. Decide whether you'll be planting directly in the ground, using raised beds or containers, or a combination of these methods. Raised beds are ideal for improving soil quality and drainage, while containers can be helpful if space is limited.

2. Sketch a rough layout of your garden area to visualize spacing, crop rotation, and companion planting strategies. This will help you make the most of your available space and resources.

3. Prepare your garden area. Once you've planned your layout, prepare the garden space. If you're using raised beds, construct or purchase frames made from untreated wood, metal, or other safe materials. For in-ground planting, clearly mark your planting rows or bed sections using stakes and string. Remove any existing grass, weeds, rocks, or debris from the area. This helps create a clean foundation and reduces competition for nutrients and water as your plants grow.

4. Prepare the soil. Aerate the soil by digging with the hoe so that it is loose rather than compact. If you live in an area that has low-nutrient soil such as sand or clay, you might consider buying some nutrient-rich soil. The goal is to provide a well-draining, nutrient-rich environment where plant roots can thrive.

5. Choose your crops. Select vegetables, herbs, and fruits that suit your climate, growing season, and family's dietary preferences. It's wise to focus on high-yield, easy-to-grow crops—such as tomatoes, beans, lettuce, carrots, zucchini, and kale—and herbs like basil or parsley. Research the best planting times for your region and aim for a mix of early, mid-season, and late-season crops to extend your harvest.

6. Gather your materials. Having everything ready ahead of time makes the planting process smoother and more efficient.

7. Plant your seeds or seedlings. Check the seed packets or plant tags for specific planting depth and spacing guidelines. Sow the seeds directly into the soil or transplant seedlings. To plant seedlings effectively, choose a well-draining soil, dig a hole slightly larger than the seedling's root ball, gently place the seedling in the hole, and firm the soil around it while ensuring the top of the root ball is level with the soil surface.

8. Water the newly planted seeds and seedlings thoroughly and then establish a schedule depending on the time of year and rainfall. Water regularly, especially during dry spells, and aim for deep watering to encourage strong root development. If you are unable to maintain a regular schedule or simply would like the convenience, establish a timed watering or dripper system to ensure your plants get the water they need.

9. Apply mulch. Once your plants are in the ground and watered, apply a 2- to 3-inch layer of organic mulch around the base of each plant. Good mulch options include straw, shredded leaves, wood chips, or grass clippings. Mulch helps retain moisture, suppress weed growth, and regulate soil temperature. Over time, it also breaks down and adds organic matter to the soil, improving its overall health.

10. Maintain your garden. Ongoing maintenance is key to a successful garden. Check regularly for weeds and remove them promptly to prevent them from stealing nutrients and water from your crops. Keep an eye out for signs of pests or diseases, and take action early using organic or natural pest control methods such as neem oil, insecticidal soap, or companion planting. As your plants grow, they may need additional nutrients—apply organic fertilizers like compost tea or fish emulsion as needed to support continued growth.

TIPS

- Plant fruits and vegetables that your family enjoys eating. There is no point in having a large harvest of vegetables that no one will eat.
- Learn the signs of ripeness for each crop and harvest regularly to encourage continued production.
- Consider incorporating perennial plants, such as fruit trees, berry bushes, and herbs, which can provide food year after year with minimal maintenance.
- Plan for staggered planting to ensure a continuous harvest throughout the growing season. This means planting crops in intervals to spread out the yield.
- At the end of the growing season, clean up your garden by removing dead plants, adding compost, and preparing the soil for the next planting cycle.

CAUTIONS

- Be aware of any allergies you or your family members may have to certain plants, pollen, or soil.
- Wear gloves and protective gear when handling soil or plants if you have sensitivities.
- Test your soil for contaminants, especially if your garden is near industrial areas or if it has been treated with pesticides. You can use a soil test kit (available online or from your local garden center) or consult local agricultural extension services.
- Monitor local weather patterns and be prepared for extreme weather events. Excessive rain, drought, or frost can impact plant growth. Consider using row covers or other protective measures when necessary.
- Store gardening tools and materials safely, especially if children or pets are present. Sharp tools and chemicals should be kept out of reach.

28 PLANT A CONTAINER HERB AND MEDICINAL GARDEN

In times of disaster, access to medicinal supplies may be limited. A container herb and medicinal garden allows you to cultivate a variety of herbs and plants that can enhance your well-being and self-sufficiency. These gardens are particularly beneficial for those with limited space who may only be able to grow their herbs indoors or on patios.

By growing herbs such as basil, oregano, and thyme, as well as medicinal plants like lavender, chamomile, and echinacea, you can create a readily available supply of ingredients for cooking, flavoring, and using as natural remedies. This not only ensures access to fresh herbs for enhancing nutrition and flavor in meals but also provides essential plant-based solutions for common ailments, stress relief, and overall wellness.

A container herb garden is also easy to manage and requires minimal upkeep, making it an accessible option for novice gardeners and those with busy lifestyles.

WHEN TO DO IT

As with your survival garden, the time to plant a container herb and medicinal garden is before an emergency takes place, as plants take time to establish. Aim to plant in spring, once the risk of frost has passed and temperatures are conducive to growth. This allows you to take full advantage of the full growing season and ensures your plants have the best chance to thrive.

MATERIALS

- Containers for planting
- Soil (high-quality potting mix)
- Seeds or seedlings
- Watering equipment
- Fertilizers
- Mulch (optional)
- Plant labels
- Pest-management supplies

TOOLS

- Trowel
- Hand fork
- Pruners
- Soil scoop

KEY ACTIONS

1. Choose a variety of herbs based on your preferences and their medicinal properties. Common options include basil, parsley, cilantro, thyme, oregano, rosemary, mint, chamomile, and echinacea. Decide which herbs and medicinal plants you want to grow based on their uses, your preferences, and your growing zone.
2. Identify a suitable space with adequate natural light, ideally receiving at least six hours of direct sunlight per day. This can be a patio, balcony, or a sunny indoor windowsill. If you're growing indoors and sunlight is limited, you may want to supplement with a grow light.
3. Select pots that are appropriately sized for the herbs you plan to grow. Choose containers made from materials like terracotta, ceramic, plastic, or metal. Ensure they have drainage holes to prevent waterlogging. Most herbs do well in 6- to 12-inch pots. If you're reusing containers, clean them thoroughly to remove any old soil or residues. Ensure that the drainage holes are unobstructed. If using a container without holes, add a layer of small stones or gravel at the bottom to improve drainage.

4. Fill each container with soil mix, leaving about 1 inch of space at the top to prevent overflow when watering. Look for mixes specifically designed for container gardening. Incorporating compost can enrich the soil with nutrients and improve its structure. Aim for a mix that retains moisture but also drains well.
5. Follow the plant's instructions for planting depth and spacing. Generally, plant seeds about 1/4 to 1/2 inch deep. If using seedlings, gently remove them from their nursery containers and plant them in the prepared soil. Ensure that the root ball is covered and the plant is at the same depth it was growing previously.
6. Water the newly planted herbs thoroughly to help settle the soil around the roots. Ensure that excess water drains out of the bottom of the containers. Moving forward, monitor soil moisture regularly and water as needed, keeping the soil consistently moist but not soggy. A comfortable watering can or a hose with a spray attachment can help manage watering effectively.
7. Consider applying organic fertilizer according to the manufacturer's instructions, usually every few weeks during the growing season. A soil moisture meter is optional but useful for monitoring levels to prevent over- or under-watering.
8. If desired, apply a layer of organic mulch on top of the soil to help retain moisture, suppress weeds, and regulate soil temperature.

TIPS

- Regularly check your herbs for growth, pests, and diseases. Remove any dead leaves or flowers to encourage healthy growth.
- As your herbs grow, prune or harvest them to promote further growth.
- If you have a bountiful harvest, consider drying or freezing excess herbs for later use. Dry herbs by hanging bundles upside down in a cool, dark place or using a dehydrator.

- Familiarize yourself with the specific growing requirements, benefits, and potential uses of each herb you plant. This knowledge will help you care for them properly and make the most of your harvest.
- Plant herbs with similar water and sunlight needs together. For example, Mediterranean herbs such as rosemary, thyme, and oregano prefer drier conditions, while basil and mint thrive in more moisture.
- If you plan to grow your herbs indoors or in a location with limited sunlight, consider using grow lights to provide adequate light for healthy growth.

CAUTIONS

- Avoid harvesting more than one-third of the plant at a time to ensure continued growth.
- Be prepared to adjust your care routines based on seasonal changes. For instance, you may need to water more frequently in hot, dry weather or move containers indoors when cool weather approaches.
- Medicinal herbs can help with some minor ailments, but they should always be used in consultation with a medical professional.

29 SPROUT NUTRIENT-RICH GREENS INDOORS

With just a few seeds, water, and a small container, anyone can cultivate a variety of sprouts, such as alfalfa, broccoli, or lentils, in a compact space. Sprouts are incredibly rich in vitamins, minerals, and antioxidants, which help support immune function and overall health, making them a valuable addition to any emergency food plan.

The act of growing our own food reminds us that we can take proactive steps to ensure our well-being. An added bonus is that these nutrient-dense greens can enhance the flavor and nutritional profile of stored foods, making meals more satisfying and enjoyable during potentially stressful times. Ultimately, sprouting greens indoors serves as a practical and sustainable solution that enhances food security and promotes health in the face of adversity.

WHEN TO DO IT

Incorporate sprouting into your emergency plan before any crisis occurs, to ensure a reliable supply of fresh food. Maintaining a regular sprouting schedule is also important, as it ensures a steady rotation of ready greens that can provide essential nutrition when fresh produce may be scarce.

MATERIALS

- Seeds for sprouting
- Glass jars, sprouting trays, or mesh sprouters
- A breathable cloth or a specialized lid for jars
- Water

KEY ACTIONS

1. Select seeds for the produce that you and your family would most like to eat.
2. If using a glass jar, ensure it is clean and dry. For trays or mesh sprouters, rinse them thoroughly to eliminate any residue.

3. Measure out your seeds. The quantity will depend on the type and size of the container you are using. Follow the package instructions to ensure the seeds have enough space to grow without becoming overcrowded. Here are some general guidelines:
    - Small container (2 to 4 inches in diameter): Use 1 to 2 teaspoons of seeds.
    - Medium container (6 to 8 inches in diameter): Use 2 to 4 tablespoons of seeds.
    - Large container (10 to 12 inches in diameter): Use 1/4 to 1/2 cup of seeds.
4. Place the seeds in the container and rinse them with clean, room-temperature water. Swirl them around gently in the water to remove any dust or debris, and discard the water.
5. Fill the container with enough water to cover the seeds by a few inches.
6. Set the container aside for six to twelve hours to allow seeds to soak. This helps to kick-start the germination process.
7. After soaking, drain the water completely. If using a jar, cover it with a breathable cloth or lid to allow airflow while preventing debris from entering.
8. Rinse the seeds thoroughly with fresh water and drain again.
9. Position the container in a warm area with indirect sunlight. Avoid direct sunlight, as it can overheat the sprouts.

10. Repeat step 8, rinsing and draining the seeds every eight to twelve hours, to keep the seeds moist but not soggy. In three to seven days, depending on seed type, you should see sprouts beginning to develop. If you notice a foul odor or fuzzy growth (mold), discard the batch and try again using fewer seeds or improving airflow.
11. Harvest sprouts when they are 1 to 2 inches long and have developed green leaves. Give them a final rinse to clean off any seed hulls.
12. Dry the sprouts by gently patting with a paper towel and store them in a resealable container in the refrigerator. They will keep for about a week.

TIPS

- If your sprouts are growing too quickly or becoming leggy (growing tall and spindly with long stems and small leaves) due to too much heat, move them to a cooler location to slow down growth and promote sturdier sprouts.
- Consistency is key to successful sprouting, so set a reminder for rinsing and draining your sprouts to ensure they stay moist and healthy.
- Different seeds may require different soaking times. Experiment to find the optimal soaking duration for each type to maximize germination rates.

CAUTIONS

- Always wash your hands and tools thoroughly before and after handling sprouts. Rinse sprouts again before eating.
- Give your seeds enough space when sprouting. Overcrowding can lead to reduced airflow and mold growth, impacting the quality of your sprouts.
- Always use filtered or clean water to rinse and soak your seeds. This helps prevent the introduction of harmful bacteria.

30 BUILD A SOLAR FOOD DEHYDRATOR

Dehydration is one of the oldest methods of food preservation, removing moisture from fruits, vegetables, and meats to inhibit the growth of bacteria and mold. Using a solar dehydrator to harness the sun's energy is a reliable way to extend the shelf life of your food, and it allows you to transform fresh produce into nutritious snacks and ingredients that can last for months. Moreover, the ability to dehydrate food allows you to leverage seasonal harvests, reducing waste and making the most of your bounty.

If you don't have access to any materials, you can still dehydrate food by slicing it thinly and laying it out in the full sun or hanging it over a low-burning fire, but you will need to watch it closely to avoid animal theft and insect contamination.

Building a solar dehydrator is a fun and rewarding project that turns food preservation into an enjoyable experience. This is not only a cost-effective way to store your harvest but also an eco-friendly method that reduces reliance on electricity. Plus, you'll gain valuable skills in food preservation while connecting with nature and appreciating the bounty of your garden or local market. Even if a disaster never happens, you'll have a ton of tasty treats to take on your next hiking trip.

There are many different types of dehydrators, but we are going to stick with the simplest design for ease of construction.

WHEN TO DO IT

Build a solar dehydrator well before any potential crisis or emergency. Ideally, construct it in late spring or early summer when many fruits and vegetables are entering peak harvest season.

MATERIALS

Frame

- 4 × (2 × 2) timber, 36 inches long
- 4 × (2 × 2) timber, 24 inches long

Legs

- 2 × (1 × 2) timber, 24 inches long
- 2 × (1 × 2) timber, 18 inches long

Plywood panels (⅜ inch thick)

- 1 × base panel, 36 × 24 inches
- 4 × wall panels (front, back, and sides), 36 × 12 inches
- Fine mesh or screen, as needed

Glazing/Top lid

- 1 × clear acrylic or polycarbonate sheet, 36 × 24 inches
- 2 × (1 × 2) timber, 36 inches long
- 2 × (1 × 2) timber, 24 inches long

Mesh tray

- 2 × (1 × 1) timber, 35 inches long
- 2 × (1 × 1) timber, 23 inches long
- Mesh or small hole wire, 35 × 23 inches

Hardware

- 2 to 3 small hinges
- 1 latch or hook
- 1 box of 1½-inch wood screws
- Staples (¼ to ⅜ inch), if using a staple gun
- Wood glue
- Weatherproof sealant

Finishing

- Matte black paint, for interior base and back panel (improves heat absorption)
- Clear exterior varnish, for weatherproofing (optional)

TOOLS

- Saw
- Drill
- Hammer
- Measuring tape
- Safety glasses
- Work gloves

KEY ACTIONS

MAKE THE FRAME (A)

1. Lay out two 36-inch pieces and two 24-inch pieces to form a rectangle.
2. Screw the corners together using 1½-inch screws, checking that the frame is square.
3. Repeat to make a second rectangle.

ATTACH AND PAINT THE BASE PANEL (B)

1. Place the 36 × 24-inch plywood panel on top of one frame to form the base.
2. Screw it down evenly along all four edges.
3. Paint the inside surface black to absorb solar heat. Allow it to dry before proceeding.

CUT AND COVER THE VENTS (C)

1. Cut a 3 × 3-inch opening low on the front panel for air intake.
2. Cut a second 3 × 3-inch opening high on the back panel for hot-air exhaust.
3. Staple or glue fine insect mesh over both openings.

ATTACH THE FRONT AND BACK PANEL TO YOUR FRAMES (D)

1. Screw the front and back panels to the frames. Install the panel with the low vent at the front and the panel with the high vent at the back.

A
B
C
D
E
F
G

ATTACH THE SIDES (E)

1. Screw one panel to each side, connecting the two frames and forming the box.

MAKE THE GLAZED LID (F)

1. Build a simple lid frame using two 36-inch and two 24-inch 1 × 2s. Place the clear acrylic or polycarbonate piece on top and secure it with screws or clips. Attach 2 to 3 small hinges along the back edge so the lid opens upward.
2. Install a latch or hook at the front to keep the lid closed.

INSTALL CORNER POSTS/LEGS (G)

1. Stand a 24-inch leg at each back corner of the base frame. Install the front legs flush with the bottom edge of the base. The back legs should extend 12 inches higher than the front legs, creating a slope (24 inches at the back, 12 inches at the front).
2. Screw each leg securely into the inside corners of the base frame.

MESH TRAY

1. Screw the four wood pieces together to form a rectangular frame.
2. Staple or screw the mesh or wire across the frame to create a removable tray that fits inside the dehydrator.

FINISHING TOUCHES

1. Seal any small gaps with weatherproof silicone to improve heat efficiency.
2. (Optional) Coat the exterior with varnish for added weather resistance.
3. Place the dehydrator on level ground or a small stand, tilted toward the sun at a 30- to 45-degree angle.
4. Arrange food on mesh tray inside.

5. Keep the lid closed while drying; open it slightly if temperatures become too high.

6. Rotate or stir food occasionally for even drying.

TIPS

- Before adding food, allow the dehydrator to preheat for about thirty minutes in direct sunlight. This helps create a warm environment for drying.
- Cut fruits and vegetables into uniform pieces to ensure even drying. Thinner slices will dry faster, while thicker pieces may need more time.
- Monitor the drying process, adjusting the position as necessary to optimize sunlight exposure.
- Once the food is dehydrated, allow it to cool before storing it in airtight containers. Glass jars, vacuum-sealed bags, or mylar bags work well. Store in a cool, dark place to extend shelf life.
- Be patient. Solar drying can take longer than conventional methods. Plan ahead and allow plenty of time for the drying process.

CAUTIONS

- Ensure that your dehydrator is well-constructed and stable. A flimsy design can collapse or tip over, especially in windy conditions.
- Avoid leaving the dehydrator exposed during rainy weather or placing it in moisture-prone areas. Moisture can lead to mold growth on the food.
- Always wash fruits and vegetables thoroughly before dehydrating. If using meat, ensure it is handled safely and cooked properly to prevent foodborne illnesses.
- Ensure that all materials used are suitable for outdoor use and can withstand the elements.

31 MAKE A DIY SOLAR OVEN

A solar oven utilizes the sun's rays to heat and cook food effectively. Using basic materials, you can design a simple and effective oven that captures sunlight, converts it into heat, and retains that warmth for cooking. Whether you're slow roasting a meal or baking bread, a solar oven can reach temperatures high enough for a variety of culinary tasks.

Solar ovens are portable and easy to use, making them ideal for families or individuals seeking self-sufficiency during crises.

There are several methods for constructing a solar oven. One is a simple, temporary design made with commonly available materials. The other is a more durable, long-term option that requires a bit more time and effort. We'll explore both methods, but I recommend including the more permanent option in your preparedness plan.

WHEN TO DO IT

Build a solar oven before an emergency. By constructing it in advance, you'll familiarize yourself with its functionality and usability, ensuring you can rely on it when needed.

TEMPORARY SOLAR OVEN

MATERIALS

- Large (2 × 2 × 1 feet) sturdy, clean cardboard box
- Aluminum foil
- 2 × 2 feet plastic wrap or plexiglass sheet
- Nontoxic black paint
- Duct tape
- Wood glue
- Dark metal pan for cooking

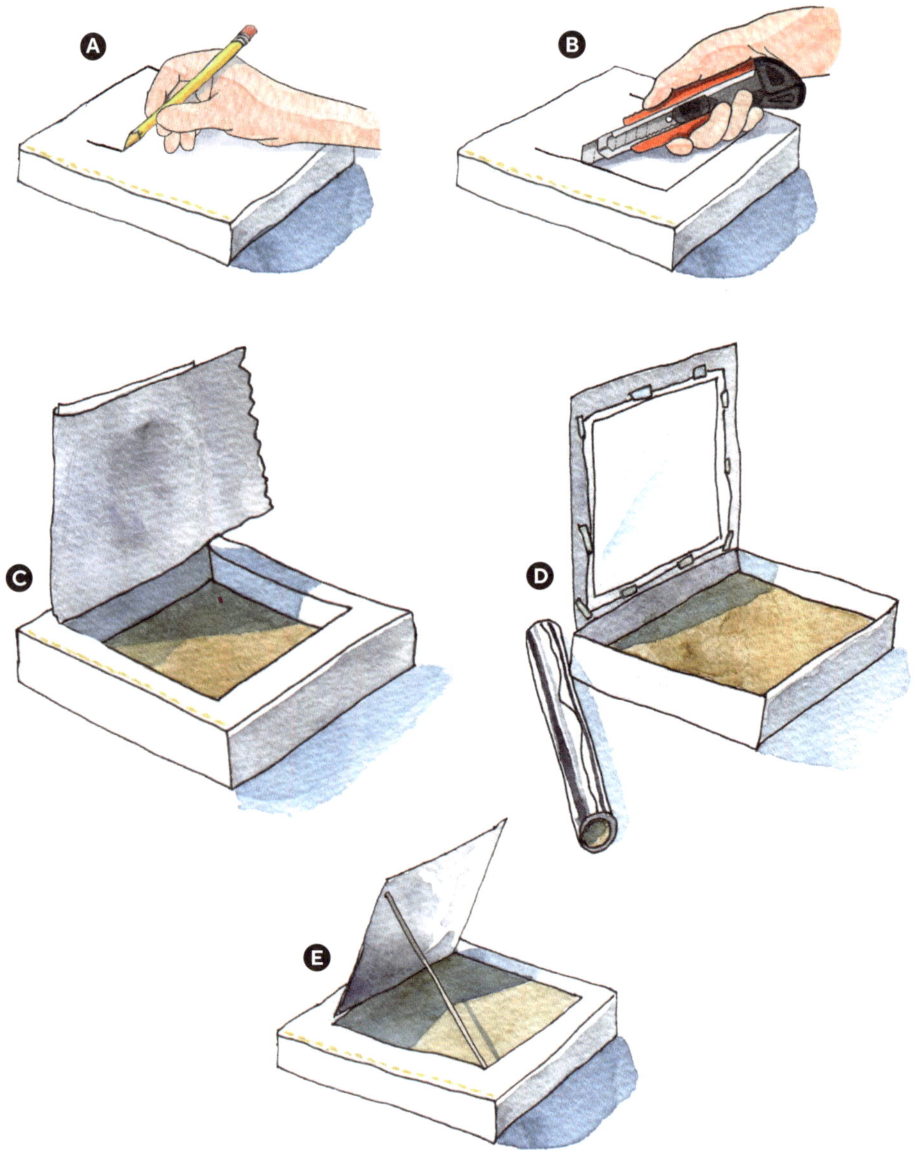
A
B
C
D
E

TOOLS

- Marker
- Scissors or utility knife
- Ruler

KEY ACTIONS

1. Draw a large rectangle on the top of the box, 1 to 2 inches from the edges. **(A)**
2. Using scissors or a utility knife, cut along three sides of the rectangle to create a flap that opens upward, like a lid. Leave one side attached (the side closest to the hinge of the box). **(B)**
3. Using the wood glue, glue the aluminum foil to the inner side of the flap, shiny side up. Smooth out wrinkles and tape it securely using the duct tape. This flap will reflect sunlight into the box. **(C)**
4. Stretch clear plastic wrap tightly over the opening of the box and tape it down around the edges to create a sealed window. **(D)**
5. Paint the inside of the box black with the nontoxic paint. This is an optional step but will increase the effectiveness of the oven.
6. Place the solar oven in a sunny location, angled so that the foil flap reflects the most sunlight into the plastic-covered opening. Use a stick or straw to prop the flap at the right angle. **(E)**
7. Preheat the oven.
8. Place food inside the dark metal pan and set it in the oven.
9. Close the lid, ensuring the plastic window is sealed, and wait for food to cook. Cooking times vary depending on the food and sunlight—monitor it regularly.

PERMANENT SOLAR OVEN

MATERIALS

- 1 base panel: 3/8-inch plywood, 2 × 2 feet
- 4 side panels: 3/8-inch plywood, 24 × 18 inches
- 1 lid panel: 3/8-inch plywood, 2 × 2 feet
- Aluminum foil
- Plexiglass sheet: 24 × 24 inches
- Dark metal cooking pan
- 2 small hinges
- 2 folding lid-support hinges
- 1 box of 1½-inch screws
- Wood glue
- 1 can matte black high-temperature paint

TOOLS

- Ruler
- Tape measure
- Marker
- Hand saw or circular saw
- Hammer
- Screwdriver

KEY ACTIONS

1. Assemble the base and walls into a box shape using screws. **(A)**
2. Line the interior surfaces (excluding the bottom) with aluminum foil, shiny side up. **(B)**
3. Attach hinges to one side of the oven for the lid. **(C)**
4. Line the lid with aluminum foil and attach to the hinges on the box.
5. Attach the folding lid-support hinges to the lid and the box to allow the lid to be supported when open.
6. Securely place the plexiglass sheet over the top of the oven. **(D)**
7. Position the solar oven in a sunny location, angled for maximum sunlight exposure.
8. Preheat the oven before placing food inside.

A

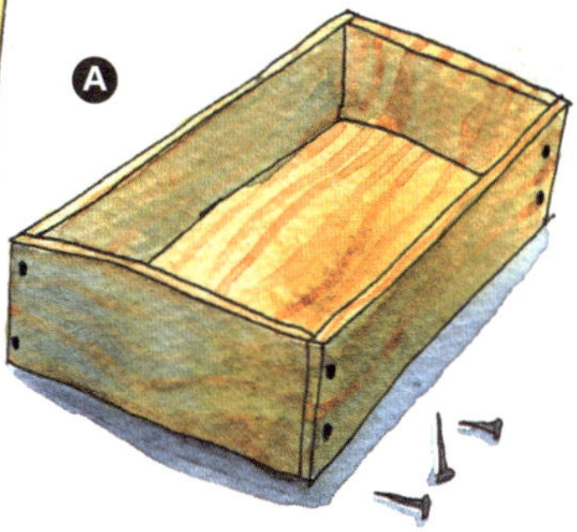

B

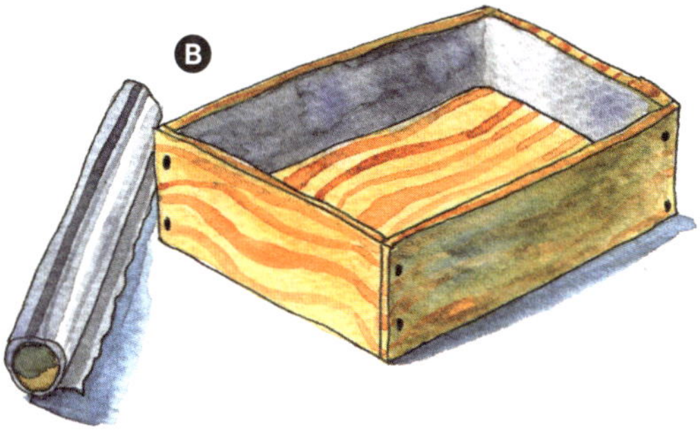

C

D

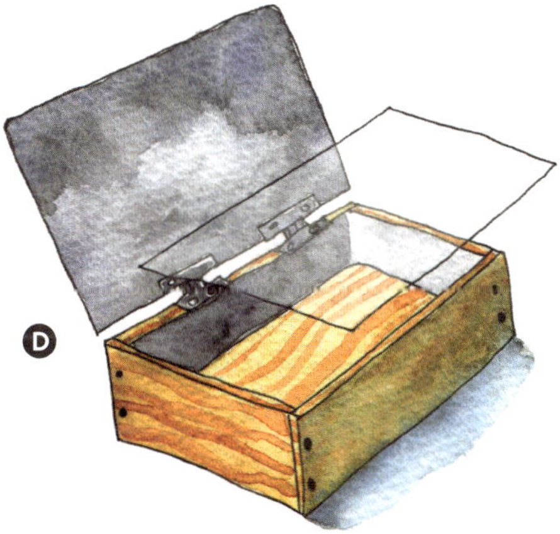

TIPS

- Use insulation such as newspaper or ash to line the inside of your oven if you wish to increase the heat. This can be done by placing a smaller cardboard box inside the larger one or lining your wooden box with plywood.
- Use your solar oven on sunny days with minimal cloud cover for the best results. The more direct sunlight, the better.
- Place the oven in a location where it will receive sunlight for the longest duration. Adjust the angle throughout the day to keep it facing the sun.
- Ensure the lid is securely closed to trap heat inside. If using a cover, make sure it's airtight to prevent heat loss.
- Smaller portions of food cook more evenly and quickly. Consider cutting larger items into smaller pieces.
- Start with simple recipes and gradually try more complex dishes as you become familiar with how your solar oven performs.

CAUTIONS

- Solar ovens can reach up to 350°F. Use oven mitts or heat-resistant gloves when handling hot containers or opening the oven.
- Make sure the structure is sturdy and well-assembled to prevent it from collapsing or blowing over in the wind.
- Keep the solar oven away from flammable materials and surfaces. Ensure that the cooking area is safe and free from combustible items.
- Always supervise the solar oven while in use, especially if children or pets are nearby. This helps prevent accidents and ensures safe operation.
- Use a thermometer to monitor the internal temperature of the oven. This helps ensure food is cooked safely and thoroughly.
- Avoid using glass containers that can shatter. Instead, use heat-resistant materials that can withstand the oven's high temperatures.

32 BUILD A PASSIVE COLD STORAGE BOX (FOR GARAGE OR BASEMENT)

Imagine being in a situation where access to fresh food is limited. A passive cold storage box can keep your food safe for several days. These boxes work by maintaining a stable temperature, providing an ideal environment for storing foods that would otherwise spoil.

A good size for a passive cold storage box is anywhere from 2 to 4 square feet, depending on the size of your family. The good news? You don't need high-tech gadgets or complex systems to build one, just some basic materials and a bit of creativity.

A passive cold storage box will help you maintain your food's freshness without relying on electricity. In the event of a power outage—whether caused by a storm, natural disaster, or other emergency—it offers a reliable way to keep perishables cool, which can make a huge difference.

WHEN TO DO IT

If you live in an area prone to natural disasters—like hurricanes, floods, or winter storms—it's wise to build a passive cold storage box in advance. When forecasts indicate a potential disaster, having it ready can ensure you're prepared for power outages that may follow.

MATERIALS

Makes one 3 x 2 x 2-foot box

- Insulation materials (rigid foam board, corkboard, or wool)
- 8 × (2 × 2) timber, 24 inches long
- 4 × (2 × 2) timber, 36 inches long
- 4 × 3/8-inch panels of plywood, 36 × 24 inches
- 2 × 3/8-inch panels of plywood, 24 × 24 inches
- 3 hinges, for the lid
- Latch to hold the lid shut
- 1 box of 1½-inch wood screws
- Weatherstripping tape
- Plastic sheet

TOOLS

- Knife
- Saw
- Measuring tape
- Screwdriver
- Hammer

KEY ACTIONS

ASSEMBLE THE OUTER BOX

1. Lay the base flat.
2. Screw the two side panels to the edges of the base.
3. Attach the front and back panels to close the box. You should now have a hollow, open-topped box.

ADD INSULATION

1. Line the inside with 2-inch-thick rigid foam board (or layers of card-board, corkboard, or wool). Cover all four walls and the base.
2. Fit insulation tightly with no gaps; fill any cracks with offcuts or sealant.

FIT THE INNER LINING

1. Cut thin plastic sheets to match the inside walls and base.
2. Affix the sheets over the insulation to create a smooth, food-safe surface. This creates a "box within a box" with insulation sandwiched in between. Thin plywood can be used instead of plastic, if preferred.

BUILD THE LID

1. Take the 36 × 24-inch top panel and add insulation to the underside.
2. Cover the insulation with your chosen lining (plywood or plastic).
3. Attach the lid with two or three hinges along the back edge.
4. Add a latch or hook to the front to hold it shut.

SEAL IT

1. Apply weatherstripping tape around the box rim to improve insulation.

TIPS

- Locate the cooler on the side of the room opposite its main entrance.
- Use a thermometer to monitor the temperature inside the box. If necessary, wrap additional insulation around the outside of the box for extra cooling, especially in warmer climates.
- Avoid opening the box unnecessarily. Keep a list of its contents and know what you need before opening to prevent losing the cooler air.
- The effectiveness of a passive cold storage box largely depends on its insulation. Invest in a high-quality insulation such as rigid foam board to prolong cooling and reduce temperature fluctuations.

CAUTIONS

- Check the internal temperature regularly with a thermometer to ensure the desired temperature for perishable items. Placing a thermometer inside is helpful.
- Avoid overloading the box, as this can impede airflow and reduce efficiency.
- Regularly clean the passive cold storage box. This prevents buildup of bacteria and odors, ensuring that food remains safe and fresh.
- Be mindful of the types of food and beverages stored. Some items, like dairy or seafood, require stricter temperature controls.

33 SET UP AN INDOOR GROW LIGHT STATION

When it comes to emergency food sources, an indoor grow light station can be a game-changer. This setup allows you to cultivate a consistent supply of nutrient-rich greens and vegetables, even in the absence of adequate natural sunlight or during times of crisis when access to fresh produce may be limited. By establishing an indoor grow light station, you can ensure a reliable source of fresh, healthy food is readily available when you need it most.

The primary advantage of this station is its ability to provide a controlled environment for plant growth. Unlike relying on unpredictable outdoor conditions, these specialized lighting systems mimic the optimal wavelengths and intensity of sunlight, enabling you to grow a variety of crops year-round, regardless of the season or weather patterns. This level of control is particularly valuable during emergencies, when self-sufficiency and the ability to sustain your household become paramount.

An indoor grow light station can be tailored to your specific needs and space constraints. Whether you have a dedicated room, a corner of your kitchen, or even a small closet, you can set up a compact and efficient system that maximizes your growing potential. This flexibility allows you to cultivate a diverse array of produce, from leafy greens and herbs to tomatoes and other vegetables, ensuring a well-rounded and nutritious food supply.

Note that this system will require electricity to operate, so make sure it is easily attached to your backup system of choice if the grid goes down.

WHEN TO DO IT

Set up your indoor grow light station during the initial stages of your emergency preparedness plan, ideally well before any potential crises.

MATERIALS

- LED or fluorescent grow lights
- Adjustable light fixtures or stands
- Growing containers (pots, trays, or hydroponic systems)
- Growing medium (soil, coco coir, peat moss)
- Seeds or seedlings
- Watering can or spray bottle
- Automatic watering system (optional)
- Timer
- Thermometer
- Hygrometer
- Gardening gloves
- Small trowel
- Fan (optional)

KEY ACTIONS

1. Assess your space and select a suitable area in your home with enough room for your setup. Ensure it has access to electricity for your grow lights and can accommodate the necessary equipment.
2. Set up your grow lights according to the manufacturer's instructions. If using adjustable fixtures, position them above your growing area, ensuring they can be easily raised or lowered as your plants grow.
3. Fill your pots or trays with the chosen growing medium, leaving some space at the top for planting seeds or placing seedlings. Make sure the containers have drainage holes to prevent waterlogging.
4. If planting seeds, follow the recommended depth and spacing guidelines on the seed packets. For seedlings, gently remove them from their original containers and place them in the prepared growing containers, ensuring the roots are well covered.
5. Use a watering can or spray bottle to moisten the growing medium. Be careful not to overwater; the soil should be damp but not soggy.
6. If using an automated watering system like drippers or sprinklers, set it up now according to the manufacturer's specifications.
7. Place a thermometer and hygrometer in the growing area to monitor temperature and humidity levels. Adjust your setup as needed to maintain optimal conditions for plant growth.

- Most indoor plants thrive in temperatures from 65°F to 75°F during the day. At night, a slight drop in temperature is acceptable, ideally around 60°F to 70°F. Some plants may have specific needs, so it's always good to check the requirements for the particular species you're growing.
- The optimal humidity level for most indoor plants typically falls between 40 and 60 percent. Higher humidity levels can benefit tropical plants, while lower humidity may suit succulents and cacti. Maintaining consistent humidity is important, as fluctuations can lead to stress or disease in plants.

8. Once everything is set up, turn on your grow lights. Most plants require twelve to sixteen hours of light per day, so consider using a timer to automate this process.
9. As your plants reach maturity, harvest the produce as needed.

TIPS

- Check your plants regularly, ensuring they receive adequate water and light. Adjust the height of the lights as the plants grow taller to maintain the right distance.

- If necessary, use a small fan to promote air circulation, which can help prevent mold and encourage healthy plant development.
- Consider starting new seeds in another set of containers to maintain a continuous supply of fresh greens and vegetables.
- Enhance the efficiency of your grow lights by incorporating reflective surfaces around your grow area. Using materials like mylar, aluminum foil, or white paint on walls can help bounce light back onto your plants.
- Keep your plants healthy and encourage better light penetration by regularly pruning them.
- Pay attention to the nutrient needs of your plants, especially when growing indoors. Use a balanced fertilizer to provide essential nutrients.

CAUTIONS

- Overwatering your plants can lead to root rot and fungal diseases. Always check the moisture level of the growing medium before watering and ensure that your containers have adequate drainage.
- If your lights are too close to the plants, they can cause light burn, leading to scorched leaves and stunted growth. Adjust the height of the lights as necessary.
- Sudden changes in temperature can stress plants. Avoid placing your grow light station near drafty windows, heating vents, or air-conditioning units. Aim to maintain a stable temperature within the ideal range for your plants.
- Indoor plants can still attract pests like aphids, spider mites, and fungus gnats. Regularly inspect for signs of infestation and address any issues promptly with appropriate treatments, such as insecticidal soap or neem oil.
- Ensure that your electrical setup is safe by using appropriate extension cords and following the manufacturer's instructions for your grow lights. Avoid overloading circuits and keep electrical components away from water to prevent short circuits or fire hazards.

PART 4

SAFETY

IN THE REALM OF EMERGENCY PREPAREDNESS, a dedicated safety plan is not just a luxury—it's a necessity. Understanding how to navigate crises effectively can mean the difference between chaos and calm, survival and vulnerability. The focus on safety helps you clear potential dangers and make informed decisions when every second counts.

One of the core principles of safety in emergencies is learning to prioritize action steps. In a crisis, the ability to assess the situation quickly and determine what needs immediate attention can be life-saving. This means knowing how to identify threats, such as natural disasters or health emergencies, and responding with a clear, thought-out strategy. By having a well-structured safety plan, you can create a sense of order allowing you to act decisively rather than reactively.

The mantra "be prepared, not scared" underscores the importance of equipping yourself with the knowledge and tools to handle emergencies. This Safety section provides essential information on first aid, home security, and personal defense, empowering you to face challenges head-on.

Understanding the balance between risk and reward is crucial in emergency situations. The instinct to act can lead to hasty decisions that may amplify danger. This section will guide you in weighing your options, teaching you to distinguish between survival needs—food, water, shelter—and wants that can wait.

Ultimately, this section is about equipping you and your family with the knowledge to navigate crises effectively. By focusing on safety, you cultivate a proactive mindset that emphasizes clear thinking, prioritization, and a balanced approach to risk. This foundation not only enhances your ability to survive but also instills confidence that you can handle whatever challenges come your way.

PASSPORT

PROJECTS

34 CREATE A FAMILY EMERGENCY BINDER

A family emergency binder serves as a centralized resource for vital documents and information to help families navigate crises with clarity and confidence. In times of stress and uncertainty, having everything organized in one place is a powerful tool.

Emergency contact lists, medical records, insurance information, important legal documents—the last thing you want to do is scramble to find them. Involving family members in the creation and organization of the binder promotes a shared understanding of your emergency plans and resources. The collective effort not only strengthens family bonds but also ensures that everyone knows where to find important information when it matters most.

Another vital benefit of the binder is its ability to foster a proactive mindset. Assembling and maintaining this resource shows a commitment to being prepared rather than reactive. This preparedness mindset can alleviate anxiety and fear, transforming the overwhelming nature of emergencies into manageable situations.

The binder can also be customized to meet your family's unique needs. Whether you have young children, elderly relatives, or pets, you can include specific plans and resources to ensure that everyone is accounted for in your emergency strategy.

A family emergency binder is more than just a collection of papers; it's a lifeline that provides clarity, promotes teamwork, and instills confidence in your family's ability to weather any storm. By prioritizing the creation and upkeep of this important resource, you're taking a significant step toward safeguarding your family's well-being in times of crisis.

WHEN TO DO IT

Create a family emergency binder well in advance of any potential crises, ideally during your emergency preparedness planning phase. Update it every few months or whenever there are significant changes in your family's circumstances, such as new medications or contact information.

MATERIALS

- Sturdy three-ring binder
- Tabs or dividers
- Clear plastic sleeves
- Paper
- Pens
- Labels
- Pockets or envelopes
- Printed template for essential contacts
- Medical information sheets
- Insurance information
- Copies of important legal documents (wills, powers of attorney, passports, driver's licenses)

KEY ACTIONS

1. Select a binder that is durable and is big enough to accommodate all of your documents. A three-ring binder is ideal, as it allows you to easily add or remove pages.
2. Decide on the key sections you want to include. Common sections are:
   - Emergency contacts
   - Medical information
   - Insurance policies
   - Legal documents
   - Emergency plans
   - Important records
3. Use dividers to separate each section. Label each divider clearly so you can quickly locate the information you need. You can use pre-made labels or create your own.
4. Fill in the emergency contacts section. Create a list that includes:
   - Family members' names and phone numbers
   - Neighbors or close friends
   - Local emergency services (fire, police, hospitals)
   - Your doctors' contact information

5. Document medical information, including:
    - Health records for each family member
    - Allergies and chronic conditions
    - Medications, dosages, and prescribing doctors
    - Vaccination records
6. Compile insurance policies, including copies of:
    - Health insurance
    - Homeowner's or renter's insurance
    - Auto insurance
    - Any other relevant policies
7. Gather legal documents, including copies of:
    - Wills and trusts
    - Power of attorney documents
    - Identification cards (driver's licenses, passports)
8. Create an emergency plan section that details:
    - Evacuation routes
    - Meeting points
    - Communication plans (how to reach each other during a crisis)
9. Include important records, such as:
    - Birth certificates
    - Social Security cards
    - Financial documents (bank account info, tax records)

TIPS

- Make sure all family members know where to find the binder.
- Review your binder regularly, ideally every few months, to ensure all information is current. Update any changes in contact information, medical conditions, or other relevant details.
- Use clear and straightforward language in your documents to ensure that everyone in the family can understand the information quickly.

- Include a table of contents at the front of your binder to make it easier to locate specific sections.
- Ensure that all documents, forms, and sections are clearly labeled. This clarity is crucial during high-stress situations when time is of the essence.
- Customize your binder based on potential local risks, such as natural disasters, and adapt plans accordingly.

CAUTIONS

- Be mindful of the sensitive information included in the binder, such as Social Security numbers, medical records, and financial details. Store the binder in a secure location (such as a fireproof safe) to prevent unauthorized access.
- While it's important to be thorough, avoid overloading the binder with unnecessary documents. Too much information can make it harder to find critical details in an emergency. Focus on essential documents and clear, concise instructions.
- Relying solely on a physical binder can be risky. Consider keeping digital copies of important documents in a secure cloud storage service to ensure access if the physical binder is lost or damaged.

35 ASSEMBLE HOME FIRST-AID KITS

A first-aid kit is an essential item for every household, go bag (see page 227), and vehicle. In situations where professional medical help is delayed or unavailable, a well-stocked first-aid kit can make a crucial difference, turning a minor injury into a manageable issue rather than allowing it to escalate into a serious health crisis.

The primary function of a first-aid kit is to provide immediate access to basic medical supplies and tools to address injuries and health issues as they arise. Having a comprehensive kit allows for prompt treatment of wounds, burns, allergic reactions, and other medical conditions. This immediate response capability can help stabilize a person's condition, alleviate pain, and prevent complications until professional medical assistance can be sought.

Note, however, that merely having the kit is not enough. Individuals must be familiar with its contents and understand how to use each item effectively. Many families find value in taking first-aid and CPR courses, which enhance their skills and preparedness even further.

First-aid kits can also be tailored to meet the specific needs of different families or individuals. For instance, a family with young children may choose to include child-specific items, while someone with known allergies might add appropriate antihistamines. This adaptability ensures that the kit remains relevant and effective in addressing the unique health needs of those it serves.

WHEN TO DO IT

Establish a first-aid kit before any potential emergency arises. Regularly update the kit based on changing family needs or local health concerns so you are always equipped to handle medical emergencies effectively.

MATERIALS

LARGE FIRST-AID KIT (TO TREAT TWO TO SIX PEOPLE)

Wound Care & Dressings

- 1–2 major wound dressings
- 2 pressure bandages
- 1–2 roller bandages (elastic or crepe)
- 2 triangular bandages
- 2 small conforming bandages
- 5 nonadherent wound dressings
- 2–4 combine dressings (large absorbent pads)
- 5–10 cotton gauze swabs (2 × 2 inches or 3 × 3 inches)
- 2 eye pads
- 10 butterfly sutures
- 20–30 adhesive dressings (e.g., Band-Aids, assorted sizes)
- 1 blister kit (moleskin + thin foam for donut dressings)

Cleansing & Antiseptics

- 10 antiseptic wipes
- 2 saline solutions (15–30 ml bottles or vials)
- 5 skin cleaning wipes
- 1 small bottle of hand sanitizer (30–50 ml)

Tapes & Fasteners

- 1 roll of hypoallergenic sports tape (1 or 2 inches wide)
- 6 safety pins

Medications & Treatments

- 10–15 pain relief tablets (e.g., ibuprofen)
- 1 tube or sachet of hydro gel (burns/cooling gel)
- 4–6 rehydration doses (e.g., Hydrolyte tablets or sachets)
- 1 lip balm (optional)

Tools & Instruments

- 1 pair of tweezers
- 2–4 splinter probes
- 1 digital thermometer
- 2–4 fever strips (optional)
- 1 small notepad and pen

Personal Protection & Hygiene

- 4 pairs of disposable gloves (nitrile or latex-free)
- 1–2 emergency blankets (thermal/Mylar)
- 2–4 sanitary items (pads or tampons, based on needs)

Emergency & Life-Saving Equipment

- 1 CPR kit (with mask or shield and instructions)
- 1 EpiPen
- 1–2 ice packs (instant or reusable)

SMALL FIRST-AID KIT (FOR CAR OR GO BAG)

- Adhesive dressings
- 2 pressure bandages
- 1 roller bandage
- 1 triangular bandage
- 5 gauze or cotton pads
- 1 strapping tape
- 10 antiseptic wipes
- 10 pain relief tablets, such as ibuprofen

KEY ACTIONS

1. Select a sturdy and portable container or bag to hold all your first-aid supplies. Ensure that it has compartments or sections for organization.
2. Collect medical supplies. Consider your family's specific needs, including any allergies or medical conditions.
3. Categorize items into groups based on their use, such as wound care, medications, and emergency tools. This organization will make it easier to find what you need in an emergency.
4. If your container has compartments, label each section clearly to identify the contents. For example, "bandages," "medications," and "emergency tools."
5. Include a first-aid manual or printed instructions for common procedures, such as CPR or how to treat specific injuries.

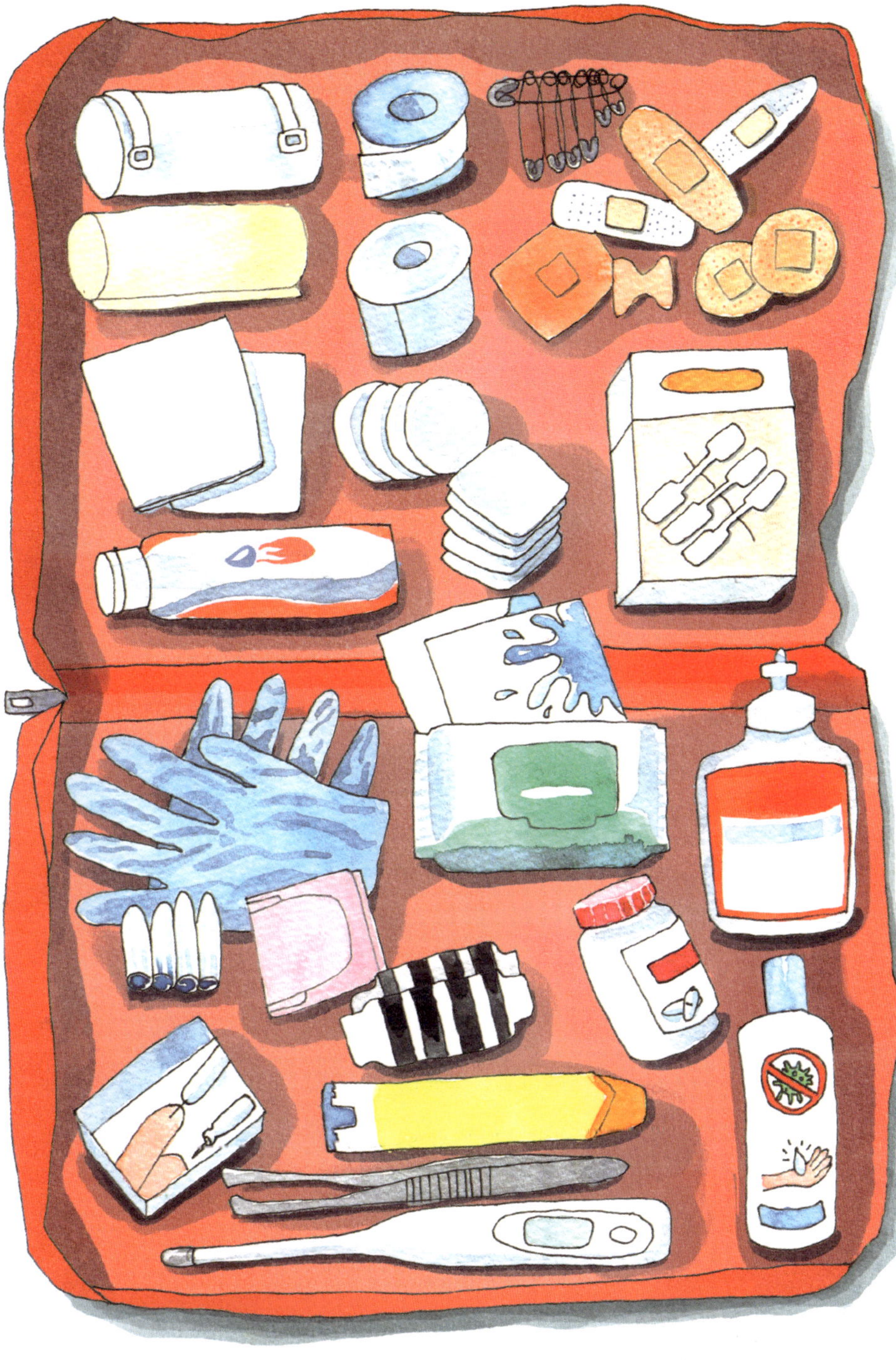

TIPS

- Store your first-aid kit in a location that is easily accessible to all family members. Ensure that everyone knows where it is kept and how to use its contents (mine is in an easily identifiable red box).
- Update the contents as needed based on family health changes, new medications, or seasonal considerations.
- Add personal items such as any prescription medications, allergy treatments, or specialized medical equipment needed by family members.
- Keep a list of important phone numbers, including local emergency services, poison control, and your family doctor, in your first-aid kit for quick reference.

CAUTIONS

- Check expiration dates and replace expired items. Using outdated supplies can lead to ineffective treatment during emergencies.
- Make sure you restock regularly if the first-aid kit is in use.
- Keep your first-aid kit in a cool, dry place away from direct sunlight and moisture. Improper storage can degrade the quality of the supplies.
- While a first-aid kit can help with many situations, it's crucial to recognize when professional medical help is needed. Don't attempt to treat serious injuries or conditions without proper training.
- Ensure that the medications and supplies in your kit do not contain allergens specific to family members. Always check labels for potential allergens.
- When using any medical supplies, carefully follow the instructions provided. Misuse can lead to complications or worsen an injury.
- Ensure that all supplies, especially those used for wound care, are sterile and stored in clean conditions to prevent infections.

36 DRAFT A FIRE ESCAPE PLAN FOR YOUR HOME

A fire escape plan serves to protect the safety and well-being of all household members. In the event of a fire, having a well-defined plan ensures that everyone is aware of the safest exit routes and any alternative paths they can take to avoid obstacles and hazards, reducing panic and confusion. It also minimizes the risk of injury or entrapment during a fire. By establishing designated meeting points outside the home, family members can quickly account for each other once they reach safety, ensuring that no one is left behind.

Practicing the fire escape plan through regular drills is essential. These practice sessions reinforce the steps that everyone should take in an emergency and allow you to identify any potential issues within the plan, such as blocked exits or confusion about meeting locations. In a real emergency, time is critical, and a practiced plan can significantly reduce response times, increasing the chances of a safe evacuation before smoke or flames become overwhelming.

Educating children about the fire escape plan empowers them to act calmly and effectively during a crisis. Their involvement not only instills a sense of responsibility but also helps them grasp the importance of fire safety. The process of creating a fire escape plan often leads to an awareness of potential fire hazards within the home, prompting necessary changes, such as removing flammable materials or ensuring that smoke detectors are functioning properly.

Make sure to include pets and domestic animals into your escape plan as well. Swift management of pets can save not only their lives but also human lives.

There's a fine line between preparing your family for an emergency and making them fearful. While it's important to ensure your family takes the drills seriously, there are also ways to make them fun. Time the event and try to beat your personal best. Do it blindfolded to simulate nighttime or smoke. Allow each participant to take responsibility for identifying their best escape route. Get creative. This way, if the worst-case scenario occurs, your family will not have to think twice to get to safety.

WHEN TO DO IT

Create a fire escape plan as soon as you move into a new home or whenever there are significant changes, such as renovations or new family members. Additionally, reviewing and practicing the plan at least twice a year helps ensure that everyone remains familiar with the procedures and any updates needed for safety.

MATERIALS

While creating a fire escape plan doesn't require specific materials, having the following items can help facilitate the process:

- Map of your home
- Paper
- Pencil
- Markers or highlighters

KEY ACTIONS

1. Examine your home layout. Identify all possible exits, including doors and windows, and ensure they are accessible and unobstructed.
2. Sketch a simple floor plan of your home on a piece of paper. Mark all exits and highlight the routes to each exit from every room.
3. Determine the quickest and safest escape routes from each room. Make sure you have several options in case your primary exit becomes blocked.
4. Designate a meeting point outside and a safe distance from your home where everyone can gather after escaping.
5. Discuss the fire escape plan with all household members, ensuring that everyone understands the routes and meeting points. Involve children in the conversation to help them feel comfortable with the plan.
6. Conduct fire drills at least twice a year to practice the escape plan. Encourage family members to take the drills seriously and follow the plan as if it were a real emergency.

7. After each drill, review the plan to identify any issues or improvements. Update the plan as necessary, especially if there are changes in your home, such as renovations or new family members.
8. Ensure that smoke detectors are installed on every level of your home, inside each bedroom, and outside of sleeping areas, and that they are in working order, as they are your early-warning system.
9. Teach family members about basic fire safety, including the importance of staying low to avoid smoke inhalation and checking doors for heat before opening. Place fire extinguishers in easily accessible locations on every floor of your home and teach all family members how to use them.

TIPS

- Keep it simple. Ensure that the escape routes and procedures are straightforward and easy to remember for all family members.
- Consider using arrows or signs to mark escape routes throughout your home, especially for children, to help guide them during an emergency.
- Test smoke detectors monthly and replace batteries at least once a year to ensure they are functioning properly.
- Include pets in the plan. Designate a person responsible for retrieving pets during an evacuation.

CAUTIONS

- Ensure that all exits are easily accessible and easily opened. Practice opening windows and doors during drills to confirm everyone knows how to do it in an emergency.
- If you have family members with disabilities or special needs, make sure your plan accommodates their requirements for safe evacuation.
- Regularly check that escape routes remain clear of any obstructions.

HOW TO USE A FIRE EXTINGUISHER

1. Pull the pin (this unlocks the extinguisher so it can be used). **(A)**
2. Aim the nozzle at the base of the fire, not the flames. **(B)**
3. Squeeze the handle to release the extinguishing agent. **(C)**
4. Sweep the nozzle from side to side, moving across the base of the fire until it is out. **(D)**

SAFETY REMINDERS

- If the fire doesn't go out quickly, evacuate immediately and call emergency services.
- Never attempt to use a fire extinguisher if you feel unsafe or if the fire is too large.

37 DEVELOP A FAMILY EMERGENCY PLAN

A family emergency plan is a key step in safeguarding the well-being of loved ones during unexpected events, particularly natural disasters. Such a plan not only helps prepare families for potential emergencies but also provides a sense of security and readiness.

When developing an emergency plan, it's important to consider the specific natural disasters most likely to affect your area. For example, families in coastal regions may need to prepare for hurricanes or flooding, while those in wildfire-prone areas should focus on creating defensible space around their homes and identifying safe evacuation routes.

Beyond addressing specific risks, an effective emergency plan includes regular practice and review. Drills help ensure that all family members, including children, are familiar with the procedures and can respond calmly if needed. It's also important to update the plan as family dynamics or living arrangements change, keeping it relevant and effective.

WHEN TO DO IT

Create a family emergency plan well in advance of potential disasters, ideally during calm periods before specific threats arise. Regular reviews and updates, at least every six months or after significant life changes, ensure that the plan remains relevant and effective for all family members.

MATERIALS

While creating a family emergency plan doesn't require specific materials, having the following items can help facilitate the process:

- Paper and pencil to sketch your home's layout, mark exit routes, and take notes during discussions
- A history of natural disasters in the area
- Local area map highlighting evacuation routes and safe locations
- Markers or highlighters to clearly indicate escape paths
- List of emergency contacts
- Go bag (see page 227)

- Emergency kit (see page 284)
- Communication device plus charging cords and power bank
- Designated meeting place information

KEY ACTIONS

1. Identify the types of natural disasters that are most likely to occur in your area, such as hurricanes, earthquakes, floods, or wildfires. Understanding these risks will help you tailor your plan.
2. Bring all family members together to discuss the importance of having an emergency plan. Encourage open communication and input from everyone.
3. Establish a system for how family members will communicate during an emergency. This should include choosing a primary and secondary contact outside the home, setting up a group chat, and ensuring everyone knows how to reach these contacts.
4. Choose two meeting locations: one close to home and another outside your neighborhood in case of evacuation. Make sure all family members know these locations.
5. Identify the safest routes to take for evacuation based on different types of disasters. Ensure that all family members are familiar with these routes and practice them if possible.
6. Create a go bag and an emergency kit that includes nonperishable food, water, first-aid supplies, flashlights, batteries, and any necessary medications. Store these supplies in an easily accessible location.
7. Keep a list of important documents and information, such as medical records, insurance policies, and contact numbers. Store this information in a waterproof container or digitally in a secure cloud service.
8. Familiarize yourself with local emergency services, shelters, and resources available in your community. This knowledge can help you make informed decisions during a crisis.

9. Conduct regular drills to practice the emergency plan, ensuring that everyone knows their roles and responsibilities. This can also help identify any areas that need improvement.

TIPS

- Ensure the plan is clear and straightforward, using easy-to-understand language.
- Create simple checklists for different scenarios to make it easier for everyone to follow.
- Store copies of the emergency plan in easily accessible locations, such as on the fridge, in a family binder, or digitally on smartphones. Make sure everyone knows where to find it.
- Consider the specific needs of all family members, including elderly relatives and those with medical conditions. Ensure the plan accommodates these needs.
- Sign up for local alert systems and follow local news channels to stay informed about potential threats.
- Set a schedule to review and update your emergency plan at least every six months or after significant changes in your family's situation, such as moving, new family members, or changes in contact information.
- Don't forget to include pets in your emergency plan. Make arrangements for their care and transportation during an emergency.

CAUTIONS

- Don't assume that everyone understands the plan or knows what to do in an emergency. Ensure that all family members are informed and trained on the plan.
- Be prepared to adapt your plan as circumstances change. Disasters can vary widely in nature and intensity, so flexibility is essential.

- Be aware of possible communication barriers, such as language differences or hearing impairments. Ensure that all family members can understand and follow the plan.
- Don't rely solely on technology. While technology can be helpful, it can also fail during emergencies. Always have a backup plan that doesn't rely on electronic devices, such as printed documents.

DISASTER-SPECIFIC CONSIDERATIONS

While it is important to be prepared for any emergency, your plan should also include specifics steps for natural disasters that are more likely to occur in your area. Here are key considerations for different types of natural disasters:

HURRICANES
- Monitor weather updates
- Evacuate early
- Secure property
- Shelter in place

EARTHQUAKES
- Drop, cover, and hold on
- Stay indoors
- Check for injuries
- Be prepared for aftershocks

FLOODS
- Evacuate if necessary
- Move to higher ground
- Stay informed
- Avoid electrical hazards

WILDFIRES
- Listen to authorities
- Evacuate promptly
- Wear protective clothing
- Stay low

TORNADOES
- Seek shelter immediately
- Protect yourself
- Stay informed
- Wait for the all-clear

38 PACK "GO BAGS" FOR EACH FAMILY MEMBER

A go bag—sometimes called a bug-out bag—is a commonly discussed component of emergency preparedness. While opinions vary widely on what to include, the core purpose remains the same: to have essential items packed and ready in case you need to leave your home quickly.

There's no one-size-fits-all list for a go bag. What it contains should reflect your environment, needs, and the most likely emergencies you might face. In general, a well-prepared go bag should help meet your basic survival needs: food, water, shelter, and warmth, along with any items that may assist with communication, hygiene, and safety.

For long-term scenarios, reusable and durable items are more effective than single-use ones. However, including such gear often requires familiarity and basic skills to use it properly—so it's important to know how everything in your bag works before an emergency arises.

That said, not every go bag needs to be packed for a doomsday scenario. In an urban setting, where evacuation due to a bushfire, flood, or power outage is more realistic, a go bag might simply include a few days' worth of essentials, such as shelf-stable snacks, bottled water, important documents, a flashlight, medications, a phone charger, and a change of clothes.

A helpful question to consider: What would you grab if you had just two minutes to evacuate your home? Some people choose to include items like photo albums, hard drives, or heirloom jewelry—anything that's irreplaceable and easy to carry. It's also wise to prepare for the possibility that emergency shelters may not be able to meet all your needs, so include items like nonperishable food, water, personal hygiene supplies, and basic clothing.

When choosing the bag itself, opt for something that's easy to carry and comfortable for extended periods, such as a sturdy daypack or backpack with padded shoulder straps. While duffel bags work in situations where you have immediate access to a vehicle, a backpack is generally more practical if walking is required.

Finally, consider making the go bag a family project. Encourage each household member to assemble their own, tailored to their age, needs, and preferences. It's a great way to start conversations about preparedness and understand what each person values most.

WHEN TO DO IT

Assemble your go bag well in advance of an emergency, ideally when creating your family emergency plan. Having it ready ensures you can quickly grab it and evacuate if a disaster strikes unexpectedly, allowing you to stay prepared for every scenario.

MATERIALS

As mentioned before, go bags are personal. Here is a list of a thorough go bag for a family of four, along with a quick explanation about each item. Feel free to add or take out different items to your own personal preference and use this as a guide rather than something set in stone. All of these things fit into a snug 40L pack.

WATER

- **Water-carrying vessel:** You can survive for two to three days without water, but you will start to lose cognitive function before then. It's important if you are going to be processing water that you have something to process and carry it in. It is my preference that this vessel is stainless steel so it can also be used as a pot to boil water if that becomes the only purification method available.
- **Water purification—straws/tablets:** These methods of purification will help make water drinkable on the go or for when making a fire to boil water may compromise your safety.

FIRE

- **All-weather lighter × 2:** Can I make fire rubbing two sticks together? Yes. Does this fire-making method burn up unnecessary time and calories? Yes. My idea of survival is making things as efficient and fast as possible, so I will most definitely include a few foolproof methods.
- **Ferro rods × 2:** Lighters will eventually run out of fuel, and have moving parts that can malfunction, so ferro rods are also a must, with the average rod offering between eight thousand and twelve thousand sparks or possible fires and being impervious to weather conditions.

SHELTER

- **Emergency blankets × 4:** These are amazing for containing body heat should temperatures plummet or water immersions occur.
- **Plastic ponchos × 4:** These are durable and will keep you dry during the wet but can also double as a waterproof roof for a shelter; they usually fold down smaller than a waterproof jacket and can be used for other purposes, such as water containment and food gathering.
- **Small tarp:** Tarps are great for shelter building, keeping in body warmth, general waterproofing, and also as a carrying tool.
- **Thermal tops and bottoms × 4:** It's always good to have warm layers for those colder times.
- **Socks × 4:** Fresh socks can aid in keeping your feet in good condition; if your feet begin to deteriorate, then your chances of survival go down dramatically.
- **Underwear × 4:** Extras are not a necessity, but luxuries are always good for survival mental health.
- **Warm hat × 4:** Contrary to popular belief, we don't lose a majority of our heat from our heads, but we do lose about 7 to 10 percent, so it's worth having a hat for cold nights.

FOOD

- **Pot/cup/mess tin × 4:** This makes dinner time a little easier; make sure it's stainless steel or similar to allow for cooking and boiling water in if necessary.
- **Ration packs for three days:** You can last twenty-one days without food, so this really is a luxury item; however, having access to something to fill your stomach until you can establish a food source can help with a good mental state of mind.
- **Spork × 4:** Building your own chopsticks is the best option for eating utensils in the outdoors, but sporks, which are a luxury item, don't weigh much or take up much room, and they make life a little easier.
- **Paracord:** The problem with cord is that it is a finite resource. For that reason, I tend not to use it for shelter building as that requires a lot of cord for a single purpose. Where possible I will use natural resources to hold my shelter together, but I will save paracord for such things as traps and snares; it is a wonderful resource for passive hunting techniques.
- **Fishing line and hooks:** The line is useful for snares and traps, and durable hooks are incredibly tricky to make out of natural materials; they are reusable, and if you are by water they are an incredibly efficient way of procuring food.

GENERAL

- **Toothbrush and toothpaste × 4**
- **Field knife:** If I am asked to choose only one survival item, then it will always be a knife. With the right ingenuity, skills, and problem-solving abilities, I can create everything I need with just a knife. This is possibly the most important item in this kit, so take your time in deciding the right knife for you. I always have a minimum of a 4-inch fixed steel, full-tang blade that I have field-tested to make sure it works well for me.

- **Multi-tool:** A good multi-tool will aid you in a variety of situations. I personally find the saw blade and screwdriver heads the most useful. This tool can help you remain adaptable both in an urban environment and in the outdoors.
- **Small first-aid kit:** I always travel with a stocked first-aid kit, and my go bag is no exception; make sure it has any prescription medication that you can't live without.
- **Spare medicines:** Medicines are best stored in a first-aid kit.
- **Hand sanitizer/gloves/masks:** These are for interaction with other people outside of your family unit, especially if your survival situation has come about during, or because of, a pandemic.
- **Head lamp × 4 :** After having spent a lot of time in the outdoors at night without a source of light, I can tell you that not being able to see is a huge safety risk and having accessible light can help hugely on a mental comfort level; it's best to have a head lamp with rechargeable batteries, but take spare batteries just in case.
- **Spare batteries:** You'll want spares in case the survival scenario becomes prolonged.
- **Flashlight:** It's always good to have a backup light source.
- **Dynamo radio/charger:** This handy gadget allows you to crank a handle to create electricity to charge it and other items; it will also give you some access to news regarding your situation if all other means of communication have failed.
- **Glow sticks:** Use these for signaling for rescue, marking locations at night, or communicating over distances.
- **Signal mirror:** Use this to signal for rescue during the day; a flash from a mirror can be seen from up to a mile away.
- **Compass:** Always valuable for wayfinding if your survival situation requires you to travel any distances, a modern compass will also usually have a magnifying glass that will aid in fire-making should the weather be sunny.

- **Duct tape:** I would use this more for repairs in an urban survival scenario, but no one can dispute the usefulness of duct tape; it is a finite resource, so use it sparingly.
- **Small solar charger:** We all have a variety of devices that require charging to function; a solar charger allows for the longevity of these items.
- **Spare cash:** If the survival scenario is urban and access to card transactions has failed, it's always good to have some cash to barter your way out of situations or buy supplies you need.
- **Photocopies of relevant documentation:** Such paperwork as birth certificates, passports, and credit cards should be photocopied and carried in the go bag in case the originals are lost and proof of identity needs to be established or money needs to be accessed.

KEY ACTIONS

1. Select a durable, lightweight backpack or duffel bag for each member of the family. Make sure it is easy to carry and large enough to hold all essential items. Keep in mind the individual who will have to carry the bag. The smaller the child, the smaller the pack.
2. Make a list for what you will include in each person's bag. Even though each pack should contain a well-rounded selection of items in case you get separated, you will need to ensure that younger or smaller individuals aren't overpacked. Choose protein bars instead of full meal packs, or one bottle of water and a LifeStraw rather than a gallon of water for the smaller packs.

   Essential items for each bag include:

   - Water
   - Food
   - Warm layers
   - Flashlight with spare batteries
   - Communication device with charger
   - Plastic poncho or emergency blanket
   - Toothbrush and toothpaste

   Larger bags can hold more items that can be shared or group items.

PASSPORT

3. Pack your go bag items. (I find that using labeled packing "cells" or smaller bags can help with organization.) Include copies of vital documents such as identification, insurance policies, medical records, and emergency contact information in a waterproof bag.
4. Store go bags in a cool, dry, and easily accessible location.

TIPS

- If you have infants, elderly relatives, or pets, include specific items such as baby formula, diapers, or pet food and supplies.
- At least every six months, review the contents of the go bag to replenish expired food and medications, update documents, and ensure the bag remains ready for use.
- Organize the items in the bag for easy access and ensure that everyone knows where the bag is stored.
- Consider labeling the contents for clarity.
- Update the bag according to the season; for example, include warm clothing and blankets in winter, and be sure to pack sunblock and insect repellent in summer.

CAUTIONS

- Avoid overpacking. Focus on essentials and prioritize lightweight, compact items.
- Ensure that everyone can carry the bag comfortably. If it's too heavy or cumbersome for certain family members, consider redistributing items or using multiple bags.
- Don't depend solely on electronic devices for communication or navigation. Include physical maps and printed emergency contacts, as technology may fail during disasters.

39 ESTABLISH A COMMUNICATION PLAN

A family communication plan is simply a pre-arranged strategy that outlines how family members will contact one another and stay connected during an emergency. It includes key information such as emergency phone numbers, designated meeting places, and backup methods of communication if traditional channels are unavailable.

In times of crisis, effective communication can be a vital lifeline. Establishing a plan before an emergency strikes helps ensure safety, coordination, and peace of mind.

During emergencies, whether natural disasters, medical incidents, or other urgent situations, confusion can set in. A clear communication plan reduces panic by making sure everyone knows how to reach one another and where to meet if separated. This sense of direction allows family members to stay safe rather than scrambling for information.

A well-thought-out plan also strengthens a family's overall resilience. It empowers individuals to respond calmly to changing circumstances and provides reassurance that there's a reliable way to reconnect, even if traditional means of communication are disrupted. This can greatly reduce stress and anxiety.

Ultimately, a family communication plan safeguards loved ones and enhances preparedness. By prioritizing communication, families are better equipped to navigate emergencies with confidence and clarity, ensuring that everyone remains connected and informed, no matter what challenges may arise.

WHEN TO DO IT

Establish a family communication plan well before an emergency occurs, ideally during times of calm, to ensure everyone is prepared and knows how to connect when it matters most.

MATERIALS

- Paper and pens
- Emergency contact information
- Local maps to identify meeting points and evacuation routes
- Sticky notes

KEY ACTIONS

1. Set a meeting time. Choose a time when all family members can be present.
2. Explain why having a communication plan is vital for safety during emergencies.
3. Identify emergency contacts. Write down important phone numbers for family members, friends, and emergency services.
4. Choose a trusted person outside the immediate area who can serve as a central point of communication. (Anyone inside the immediate area is probably also compromised by the disaster.)
5. Identify one location close to home and another farther away (about a mile away) where family members can gather if separated. Use local maps to highlight these meeting points for easy reference.
6. Talk about how your family will communicate during an emergency (e.g., cell phones, text messaging, social media). Plan for alternative ways to connect if primary methods fail, such as landlines or in-person meetings.
7. Document the communication plan, including emergency contacts, meeting locations, and communication methods.
8. Make sure each family member has a copy of the plan, both digitally and in print.
9. Post the communication plan in a visible area of your home for quick reference.

10. Regularly practice the communication plan to ensure everyone knows what to do and where to go during an emergency.

TIPS

- Ensure that all family members understand the terms and instructions in the communication plan. Avoid jargon and make it as straightforward as possible.
- Assign specific roles to family members in the event of an emergency. This could include who is responsible for checking in with the designated contact or gathering supplies.
- Create a simple code word or phrase that can be used to signal safety or distress. This can help convey important messages quickly and discreetly.
- In addition to cell phones, consider having walkie-talkies or other communication devices available, especially if you live in an area prone to power outages or disruptions.

CAUTIONS

- While it's important to have a comprehensive communication plan, too much information can lead to confusion. Keep the plan concise and focused on essential details.
- Be cautious about sharing sensitive personal information in the communication plan. Limit details to only what is necessary and ensure that contact lists are kept private.
- While preparedness is important, avoid creating a culture of fear. Frame discussions around emergency planning positively, focusing on empowerment and resilience rather than fear of the unknown.

40 DESIGN A KID-FRIENDLY COMFORT KIT

A kid-friendly comfort kit prioritizes the emotional and psychological well-being of children during stressful situations. Emergencies, such as natural disasters, evacuations, or other unexpected events, can be frightening and overwhelming for kids. A comfort kit can offer them a sense of security and calm.

Children often struggle to grasp the full scope of emergencies, which can heighten fear and anxiety. Including favorite toys, books, or blankets offers elements of normalcy during uncertain times.

Beyond comfort, the kit can also serve a practical purpose. Items like a flashlight, water, nonperishable snacks, and a small first-aid kit ensure that children have access to basic necessities if they need to evacuate or remain in a safe place for an extended period. Being prepared in this way can help them feel more in control.

Involving children in creating their own comfort kit can be a valuable bonding experience. It gives them a voice in the process and fosters a sense of ownership and responsibility. Teaching kids about emergencies in a calm and constructive manner can also help them build resilience and problem-solving skills.

By incorporating a kid-friendly comfort kit into your overall emergency preparedness plan, you help ensure your children will feel safe, supported, and ready to face any challenges that may arise.

WHEN TO DO IT

Assemble a kid-friendly comfort kit well before any potential emergencies occur. Regularly update the kit to reflect your children's changing needs and preferences.

MATERIALS

The items in your kid-friendly comfort kit will vary on the child and their age and interests. I like to include the following basics.

Emotional Comfort & Entertainment

- Small toys or puzzles
- Favorite blanket or stuffed animal
- Books or coloring materials
- Comforting notes or drawings
- Family photos
- Music

Food & Hydration

- Snacks (nonperishable)
- A small bottle of water

Hygiene & Cleanliness

- Wipes
- Hand sanitizer
- Toothbrush/toothpaste
- Spare diapers and wipes (if necessary)

Clothing & Physical Comfort

- Warmer layer of clothing
- Pajamas

Health & Safety

- Home first-aid kit (see page 214)
- Emergency contact information

Practical Items

- Flashlight or headlamp

KEY ACTIONS

1. Sit down with each child and talk about the project. Together, make a list of items to include in their kit.
2. Choose a sturdy, waterproof bag or container to hold the kit.
3. Organize all items neatly in the bag or container.
4. Seal the bag and store it in a designated, easily accessible location.

TIPS

- Schedule regular reviews of the kit every six months or annually to update items as your child grows or preferences change.
- Opt for clear bags or containers so your child can see the contents, making it easier for them to find what they need in a stressful situation.
- Consider various emergency situations your family might face and include items that are useful for those scenarios, such as a small flashlight or whistle if you are worried about electricity going out or getting lost in the woods.
- Teach your child about the purpose of the kit and how to use its contents.
- Simulate emergency scenarios where your child can practice using items from the kit, reinforcing their understanding of its importance.

CAUTIONS

- Choose lightweight and compact items to ensure the kit is easy for your child to carry or for you to transport if needed.
- Avoid the temptation to let your child play with the kit when it's not necessary, as this may result in missing items when needed.

41 PLAN FOR PET SAFETY

Pets, like humans, can be at significant risk during natural disasters or other crises. A comprehensive emergency plan that includes pets can help protect them from harm, minimize their chances of getting lost or injured, and ensure their health and well-being are prioritized.

When faced with the possibility of evacuation, a well-prepared plan also allows pet owners to quickly gather their animals, necessary supplies, and documentation. This efficiency can be vital in chaotic circumstances, ensuring that pets are not left behind. It's essential to include provisions for their needs, such as food, medications, and comfort items, which helps alleviate stress and maintain their health during a crisis.

Emergencies often lead to situations where pets may become separated from their owners. A proactive plan that includes identification methods—like microchips or ID tags—can significantly reduce the risk of losing pets in the confusion that can arise during an emergency.

Never assume you can return home to retrieve your pets during an emergency. Always include them in your evacuation plans to avoid leaving them in potentially dangerous situations. A little bit of foresight can save valuable time and reduce stress when it matters most.

Knowing that there is a solid plan in place for pets provides peace of mind. It allows you to focus on other critical aspects of emergency preparedness, reducing anxiety and fostering a greater sense of security. Responsible pet ownership includes ensuring that pets do not become a burden on emergency services or add to the chaos during a crisis.

WHEN TO DO IT

Develop an emergency plan for your pets as part of your overall emergency preparedness efforts, ideally before any potential threats arise. Regularly review and update the plan, especially after significant changes in your household or community.

MATERIALS

- Identification items (ID tags with your contact information)
- Pet carrier or crate
- Nonperishable pet food
- Portable water bowl
- Fresh water (enough to last several days)
- Medications and medical records
- Familiar items like toys, blankets, or bedding
- Leashes and collars
- List of emergency contacts (veterinarian, local animal shelters, and pet-friendly hotels)
- Waterproof bag or container

KEY ACTIONS

1. Evaluate your pet's specific needs, including dietary requirements, medical conditions, and any special accommodations they may need during an emergency.
2. Ensure your pet has proper identification, including ID tags with your contact information. Consider microchipping your pet for added security.
3. Obtain a sturdy, comfortable pet carrier or crate suitable for your pet's size. Make sure it is well ventilated and easy to transport.
4. Gather copies of your pet's medical records, including vaccination history and any necessary medications. Keep these documents in a waterproof folder.
5. Pack nonperishable pet food, a water bowl, and water for your pet. Also include familiar items like toys, blankets, or bedding to help reduce stress.
6. Ensure you have sturdy leashes and collars for each pet.
7. Make a list of emergency contacts, including your veterinarian, local animal shelters, and pet-friendly hotels. Keep this list readily accessible.
8. Familiarize yourself with local evacuation routes and identify pet-friendly shelters. Print maps highlighting these routes and keep them in your emergency kit.

9. Store all the gathered materials in a waterproof bag or container, keeping everything organized and protected for easy access during an emergency.

TIPS

- Regularly review the emergency plan with your family and practice how to quickly gather your pets and supplies. Ensure everyone knows the plan and where to find the supplies.
- Ensure your pets are comfortable with their carriers by practicing crate training well in advance. This can help reduce anxiety during an emergency evacuation.
- Expose your pets to various emergency sounds (like loud alarms) in a controlled setting to help them acclimate and reduce fear during an actual event.
- Collaborate with friends who have pets to create a support system. If you are unable to evacuate, they can assist with your pets.
- Reinforce training commands that will be useful during emergencies, such as "come," "stay," and "leave it."

CAUTIONS

- In emergencies, pets can sense human anxiety. Stay calm and composed while handling your pets to help them feel more secure.
- Always use secure leashes and collars when transporting your pets in emergency situations. A frightened pet may try to escape.
- Verify in advance which emergency shelters are pet-friendly and what documentation may be required.
- When evacuating or seeking shelter, be aware of other animals that may be present, as this can lead to stress or aggressive behavior in pets.

42 OBTAIN A NOAA WEATHER RADIO RECEIVER

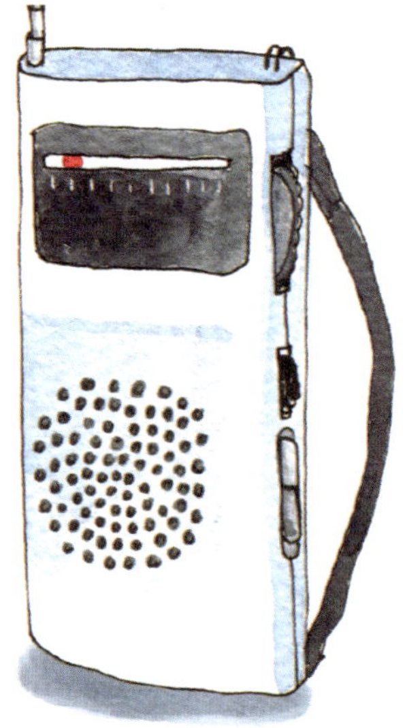

A NOAA Weather Radio receiver is a specially tuned radio that picks up weather information and alerts on specific VHF frequencies from the National Weather Service (NWS). These receivers provide 24/7 broadcasts of weather forecasts and hazard alerts, such as thunderstorms, tornadoes, hurricanes, and floods. Receivers come in various forms, from simple portable units to those with advanced features like Specific Area Message Encoding (SAME) for localized alerts, battery backup, and even solar or hand-crank power for emergencies, ensuring that crucial information is accessible when it's needed most. One of the key features of NOAA Weather Radio is its ability to transmit alerts and warnings even during power outages. Many NOAA Weather Radios come equipped with battery backup systems, ensuring that crucial information is accessible when it is needed most. This feature is especially important during severe weather events when traditional communication methods may fail.

The National Oceanic and Atmospheric Administration (NOAA) broadcasts continuous weather information and emergency alerts from the NWS through a network of radio stations across the United States. These broadcasts are designed to keep the public informed about weather conditions, forecasts, and warnings. They also include detailed information about the nature of the weather threat, expected impacts, and safety recommendations. The information is delivered in an easily understandable format, making it accessible to everyone.

In addition to weather alerts, NOAA Weather Radio also provides routine weather forecasts, marine weather reports, and information about hazardous weather conditions, ensuring that listeners are well-informed about their local weather. The broadcasts are made in multiple formats, including voice and tone alerts, which can be particularly helpful for individuals with hearing impairments.

WHEN TO DO IT

Purchase a NOAA Weather Radio in advance of any potential crises, and tune into NOAA broadcasts early on, as you prepare for potential emergencies. It is also advisable to check radio communications whenever there are forecasts of severe weather, such as storms, hurricanes, or flooding, to receive timely alerts and updates.

MATERIALS

- NOAA Weather Radio receiver, with user's manual
- Antenna (if needed)
- Power source

KEY ACTIONS

1. Carefully unpack your NOAA Weather Radio and ensure all components are included, such as the radio unit, power adapter, and user's manual.
2. Choose a suitable location for the radio. Ideally, it should be placed near a window or in an elevated area to improve signal reception. Avoid areas with heavy electronic interference such as near microwave ovens or computers, which may disrupt signals.
3. If your radio is AC-powered, plug the power adapter into an electrical outlet. If it runs on batteries, insert fresh batteries according to the instructions in the user's manual.
4. If your radio has an external antenna, extend it fully to enhance reception. Ensure it is positioned vertically for optimal performance.
5. Power on the radio. You should see lights indicating it is operational.
6. Use the tuning dial or buttons to set the radio to the appropriate NOAA frequency for your area. Find local frequencies on the NOAA website or

in the user's manual. Many radios have a "Weather" or "Alert" button that automatically tunes to the correct frequency.

TIPS

- If your radio has the capability to program specific alerts, follow the instructions in the user's manual to set your location so you receive localized alerts relevant to your area.
- Listen to the broadcast to ensure the radio is receiving signals clearly. Adjust the antenna or the radio's position if the signal is weak.
- Take time to explore additional features such as alert tones, volume control, and any other functions your radio may have, as outlined in the user's manual.
- Periodically check the batteries and replace them as needed, especially before severe weather seasons, and ensure the radio remains operational.
- Familiarize yourself with the types of alerts and warnings broadcasted by NOAA. Misunderstanding these alerts could lead to inadequate responses during emergencies.

CAUTIONS

- Use the NOAA Weather Radio as part of a broader emergency preparedness plan. Don't rely solely on it for information—supplement it with other sources like local news, weather apps, and alerts.
- During severe weather events, keep the radio on to receive continuous updates. Conditions can change rapidly, and staying informed is crucial for safety.
- Do not attempt to modify the radio or its antenna. Unauthorized alterations can compromise the radio's performance and may violate warranty terms.

43 INSTALL AND MAINTAIN HOME SAFETY DEVICES

Smoke alarms, carbon monoxide (CO) alarms, and fire extinguishers play a vital role in protecting households from the devastating effects of fires and related hazards. They are considered the first line of defense and are essential for ensuring the safety of occupants.

Smoke alarms are crucial for early detection of fires. By sensing smoke and sounding an alarm, they provide critical time for individuals to evacuate safely. Installing smoke alarms in key areas of the home, such as bedrooms, hallways, and living spaces, significantly enhances the chances of survival in the event of a fire. Regular maintenance is necessary, including testing alarms monthly and replacing batteries at least once a year, to ensure they function properly.

Similarly, CO alarms are indispensable for detecting carbon monoxide, a colorless and odorless gas that can be emitted by malfunctioning fuel-burning appliances, fireplaces, or vehicles. CO is highly toxic and can lead to serious health issues or even death if not addressed promptly. Installing CO alarms near sleeping areas and on every level of the home helps protect occupants from the dangers of CO poisoning.

Fire extinguishers serve as a critical tool for controlling small fires before they escalate into larger, more dangerous situations. Having a fire extinguisher readily accessible—especially in areas prone to fire hazards, such as the kitchen, garage, and near heating equipment—empowers individuals to take immediate action in case of a fire. Every household member should know how to use a fire extinguisher effectively (follow the PASS technique: Pull, Aim, Squeeze, Sweep).

Together, smoke alarms, CO alarms, and fire extinguishers create an integrated safety system that enhances household fire preparedness. They provide early warning and a means of control, and they promote a culture of safety, encouraging occupants to be proactive about fire risks and emergency response. Regularly maintaining these devices and fostering awareness about their importance can significantly reduce the likelihood of fire-related injuries and fatalities in the home.

WHEN TO DO IT

Smoke alarms, carbon monoxide (CO) alarms, and fire extinguishers should be standard in every household. Regular checks and maintenance of these safety devices should be conducted at least once a year, but it's best to assess them more frequently to ensure they are always ready for emergencies.

SMOKE ALARMS

MATERIALS

- Smoke alarms (battery-operated or hardwired)
- Batteries (if applicable)
- Screws or mounting hardware (usually included with the alarm)
- Screwdriver
- Ladder or step stool

KEY ACTIONS

1. Select appropriate locations for the smoke alarms. You should have one in every bedroom, outside each sleeping area, and on every level of the home, including the basement.
2. Use a ladder or step stool to reach the installation points safely. Clear any obstructions around the installation area.
3. Insert new batteries, if needed.
4. Following the manufacturer's instructions, secure the alarm to the ceiling or wall.
5. Once installed, press the test button on the smoke alarm to ensure it is functioning correctly. You should hear a loud beep if the alarm is operational.

CARBON MONOXIDE (CO) ALARMS

MATERIALS

- Carbon monoxide alarms (battery-operated or hardwired)
- Batteries (if applicable)
- Screws or mounting hardware (usually included with the alarm)
- Screwdriver
- Ladder or step stool

KEY ACTIONS

1. Choose installation locations for the CO alarms. Install them near sleeping areas and on every level of the home, particularly near potential sources of CO, such as kitchens or garages.
2. Use a ladder or step stool to reach the installation points safely. Clear any obstructions around the installation area.
3. Insert new batteries, if needed.
4. Following the manufacturer's instructions, secure the alarm to the ceiling or wall.
5. Once installed, press the test button on the CO alarm to verify it is working properly. A beep or light indicator should confirm functionality.

FIRE EXINGUISHERS

MATERIALS

- Fire extinguishers rated for household use (Class A, B, or C ratings, based on the types of fires you may encounter)
  - Class A: ordinary combustibles
  - Class B: flammable liquids
  - Class C: electrical fires
- Mounting brackets
- Screws
- Screwdriver

KEY ACTIONS

1. Determine the best locations for fire extinguishers. Common areas include the kitchen, garage, and near heating equipment. Ensure they are easily accessible.
2. Use a mounting bracket to secure the fire extinguisher to the wall. Follow the manufacturer's instructions for proper height and placement—typically, the handle should be 3 to 5 feet above the ground.
3. Ensure the fire extinguisher is fully charged and that the pressure gauge indicates it is in the green zone. Remove any plastic covers or seals.
4. Make sure everyone in the household knows the location of the fire extinguishers and how to operate them effectively using the PASS technique (Pull, Aim, Squeeze, Sweep).

TIPS

- When buying a smoke alarm, opt for dual-sensor models that detect both ionization and photoelectric smoke. This ensures better detection of different types of fires.

- When buying carbon monoxide alarms, choose models that are UL-listed and have a digital display for easy monitoring.
- Decide between battery-operated and hardwired models. Hardwired alarms are connected to the home's electrical system and often have backup batteries for reliability. They will need a professional to install. Battery-operated models offer flexibility in placement.
- Select extinguishers that are easy to handle. A 2.5- to 5-pound extinguisher is typically manageable for most adults.
- Ensure fire extinguishers are mounted in easily accessible locations, free from obstructions, and visible in case of an emergency.
- Schedule regular maintenance checks for all alarms and extinguishers. Test smoke and CO alarms monthly, replace batteries annually, and check the pressure gauge on fire extinguishers to ensure they are charged.

CAUTIONS

- Ensure all alarms and extinguishers are certified by relevant safety organizations, such as Underwriters Laboratories (UL) or the National Fire Protection Association (NFPA).
- Do not install smoke alarms in areas with high humidity, such as bathrooms, as steam can trigger false alarms.
- Install smoke and CO alarms on ceilings or high on walls, as smoke and gases rise. Ensure there are no obstructions within 12 inches of the alarm.

44 REINFORCE DOORS AND WINDOWS

Fortifying doors and windows enhances the security of a home against potential threats, whether from natural disasters, break-ins, or other unforeseen events. Because such threats can arise unexpectedly, taking proactive measures to strengthen these entry points can significantly improve safety and provide peace of mind.

Doors serve as the primary barrier between the interior of a home and the outside world. Many residential doors, especially hollow-core or lightweight ones, are vulnerable to forced entry. Upgrading to solid wood or reinforced steel doors offers a much stronger defense. It's equally important to ensure that the doorframe is secure and sturdy, which may involve installing longer screws in the hinges and strike plates to better withstand impact.

Dead bolt locks are another key component of door security. While a single-cylinder dead bolt is common, a double-cylinder dead bolt, which requires a key on both sides, can offer added protection, particularly if there is a window nearby. However, it's important to weigh this option carefully, as this can pose a safety hazard during emergencies when a quick exit is necessary.

Windows, often considered weak points, also deserve attention. Reinforcing the glass with security film, which is a thin layer of polyester, can make the glass more difficult to shatter. Alternatively, tempered glass, which is designed to be more resistant to impact, can also deter break-ins. For added protection, window locks and security screens not only prevent unauthorized entry but also serve as a visual deterrent.

In addition to physical reinforcements, having an emergency plan that includes procedures for quickly securing these entry points can be invaluable. Regularly reviewing and practicing this plan ensures that all household members are prepared to act in a crisis.

If you are especially concerned for your safety, a sensor and alarm system can offer peace of mind. However, these systems are costly and most of them require professional installation, so we are going to focus on practical, budget-friendly steps you can do independently.

We'll look at fortifying entranceways against both natural disasters and human threats. You don't need to apply every method; just taking one or two steps can make a difference. Choose the options that feel right for your needs and address your concerns.

Ultimately, fortifying doors and windows is not just about physical security; it's about creating a safe environment where individuals and families feel secure. By upgrading materials, improving locks, and preparing a comprehensive emergency plan, you can greatly increase your home's resilience against potential threats.

NOTE: Strengthening community ties can enhance security. Collaborate with neighbors to keep an eye on each other's properties, especially during vacations or when severe weather is predicted.

WHEN TO DO IT

Fortify doors and windows before known threats like natural disasters, during home renovations, or early on in your preparedness plan. Regular maintenance checks should occur during seasonal changes or before extended absences.

WINDOWS

MATERIALS

- Security film
- Window bars or grilles
- Reinforced window frames
- Plywood
- Storm shutters
- High-quality locks
- Latches and bolts
- Window security screens
- Caulking
- Alarm sensors

TOOLS

- Hammer
- Screwdriver set (flathead and Phillips)
- Tape measure
- Level
- Pliers (needle nose and standard)
- Wrench set (adjustable and fixed)
- Utility knife
- Drill (manual or hand-powered)
- Saw (hand saw or hacksaw)
- Pry bar
- Allen wrench set
- Claw hammer
- Safety goggles
- Work gloves

KEY ACTIONS

1. Inspect each window to identify weaknesses, such as old glass, flimsy frames, or inadequate locks.
2. Based on your assessment, select the appropriate materials for fortification, such as security film or window bars. If you need the glass replaced, it's best to hire a professional to do the job. Window fitting is a very specialized skill.
3. Install security film (if using):
   - Thoroughly clean the window glass to remove dust and debris.
   - Measure the glass pane and cut the security film to size, leaving a little excess around the edges.
   - Peel off the backing and carefully apply the film to the glass, using a squeegee to smooth out bubbles and ensure a firm bond.
4. Install window bars or grilles (if using):
   - Measure the dimensions where the bars will be installed.
   - Cut metal bars to fit the window dimensions.
   - Use brackets and screws to attach the bars securely to the window frame.

5. If your frames are weak, consider reinforcing them with additional materials:
    - Carefully remove any decorative trim around the window.
    - Use plywood or metal brackets to strengthen the frame, securing them with screws.
    - Once reinforced, reattach the trim to hide the modifications.
6. Upgrade locks:
    - Unscrew the existing locks and remove them from the window.
    - Follow the manufacturer's instructions to install high-quality locks, ensuring they fit snugly and operate smoothly.
7. Use caulking or weatherstripping to seal any gaps around the window to prevent drafts and enhance strength.
8. Install window security screens. These are made of strong mesh and can prevent intruders from accessing windows, while still allowing ventilation and visibility.
    - Measure the dimensions of your window frame to ensure you purchase the correct size security screens.
    - Clean the window frame thoroughly to remove any dirt or debris that could affect the installation.
    - Depending on the type of security screen, choose the appropriate mounting method. If using a fixed screen, align it with the frame and use screws or anchors to secure it in place, ensuring it fits snugly. For retractable screens, follow the manufacturer's instructions to install the track or mounting brackets.
    - Once the screen is properly aligned, secure it according to the installation guidelines. Ensure that all screws are tight and that the screen is firmly attached to the window frame.
    - Test the screen for stability and make any necessary adjustments to ensure it operates smoothly.

9. Install storm shutters:

    - Select the appropriate storm shutter material and style for your windows, such as roll-up, accordion, or panel shutters. Ensure they are the right size and fit for your windows.
    - Clean the window frames and remove any obstacles or debris. Measure the window dimensions accurately to ensure that the storm shutters can be mounted securely.
    - Follow the manufacturer's instructions to attach the mounting brackets or tracks to the wall or window frame.
    - Use a level to ensure proper alignment and secure the hardware with screws or anchors as needed.
    - Once the hardware is in place, attach the storm shutters according to the provided guidelines, ensuring they open and close smoothly.

10. If desired, install alarm sensors that connect to your home security system, placing them near the window to detect unauthorized entry (see page 260).
11. Test all locks, ensure that the security film is properly adhered, and verify that windows can still be opened and closed as needed.

TIPS

- Decorative grilles can enhance the aesthetic appeal of your windows while providing added protection.
- Trim shrubs and trees near windows to eliminate hiding spots for potential intruders. Keeping your landscaping tidy can deter break-ins and improve visibility.

CAUTIONS

- Before making modifications, ensure that your fortification methods comply with local building codes and regulations. Some areas may have specific requirements for window security features.
- When installing security features like bars or grilles, make sure they do not obstruct emergency exits or escape routes. It's essential that all occupants can exit safely in case of an emergency.
- While it's crucial to enhance security, overengineering windows with excessive reinforcements may create issues with ventilation, light, or accessibility. Balance security with practicality.

DOORS

MATERIALS

- Solid-core or metal doors
- Dead bolt locks
- Door-reinforcement plates
- Security screen
- Doorjamb-reinforcement kits
- Hinge reinforcements
- Chains
- Security cameras or doorbell cameras
- Alarm system sensors
- Weatherstripping

TOOLS

- Screwdriver set (flathead and Phillips)
- Drill
- Tape measure
- Level
- Hammer
- Chisel
- Pry bar
- Pliers
- Utility knife
- Claw hammer
- Allen wrench set
- Safety goggles
- Work gloves

KEY ACTIONS

1. Inspect your existing door for weaknesses, including the type of door, locks, and frame. Determine what improvements are needed. Based on your assessment do any or all of the following.
2. Replace the door:
   - Unscrew the hinges and carefully remove the door from the frame.
   - Place the new solid-core or metal door in the frame, ensuring it fits snugly. Secure the hinges to the door first, then attach them to the doorframe.
3. Install the dead bolt lock:
   - Measure and mark the desired height for the dead bolt on the door.
   - Use a hole saw to drill a hole for the dead bolt and a smaller hole for the latch.
   - Insert the dead bolt into the hole and secure it with screws, ensuring it operates smoothly.
4. Attach reinforcement plates to the doorjamb around the lock and hinges using screws for added strength.
5. Replace the existing hinges with reinforced hinges or install hinge bolts that secure the door to the frame, preventing it from being removed.
6. Add security screen:
   - Measure the dimensions of your doorframe to ensure you purchase the correct size security screen.
   - Remove any existing door hardware or obstacles that may interfere with the installation.
   - Clean the doorframe thoroughly.
   - Align the frame with the doorframe, ensuring a snug fit.

- Use a level to check for alignment and then secure the frame in place using screws or anchors, making sure it is firmly attached.

7. Install sensors or cameras according to the manufacturer's guidelines, ensuring they are functional and properly connected to your alarm system (see page 260).
8. Apply weatherstripping around the edges of the doorframe to seal gaps, improving energy efficiency and adding an extra layer of security.
9. After completing the installation, test the dead bolt, locks, and any security devices to ensure they function properly. Make any necessary adjustments.

TIPS

- Weigh the cost of fortification materials and installation against the level of security they provide. Some features may not offer a significant improvement relative to their cost.
- While fortifying your doors, choose materials and designs that maintain the visual appeal of your home.

CAUTIONS

- Before making any modifications, ensure that your fortification methods comply with local building codes and regulations.
- Ensure that your security enhancements do not obstruct emergency exits or escape routes. Occupants should be able to exit safely in an emergency.
- While it's essential to enhance security, avoid overengineering the door, which may hinder accessibility for family members or emergency responders.
- Over-reinforcing a door can lead to issues with door operation, such as difficulty opening or closing. Ensure that all added features work smoothly together.

45 SET UP LOW-TECH ALARM SYSTEMS

Low-tech alarm systems offer a reliable and cost-effective way to enhance security without the complexity and expense associated with high-tech solutions. These systems often rely on simple mechanical devices or straightforward setups that do not require advanced technology or internet connectivity, making them accessible to a wide range of users. For example, door and window alarms that use basic sensors can be installed with minimal tools, alerting homeowners when a door or window is opened unexpectedly.

Another reason to consider low-tech alarm systems is their reliability. Since they do not rely on electricity or internet connectivity, they are less susceptible to power outages or hacking attempts, which can compromise more advanced systems. For example, a simple bell or mechanical chime can be used to alert homeowners when someone enters or exits a space, providing a consistent and dependable form of notification.

Additionally, many of these devices are available at a fraction of the cost, making them accessible to those on a budget. Some popular examples of low-tech alarm systems include:

- Door and window alarms
- Motion-activated lights
- Basic chime or bell systems

Low-tech alarm systems provide a straightforward and reliable approach to enhancing home security. Their ease of installation, affordability, and resilience to outages make them an appealing choice for homeowners seeking effective security solutions without the complications of high-tech systems. By incorporating various low-tech devices into your home, you can create a customized and robust security setup tailored to your specific needs.

WHEN TO DO IT

Set up low-tech alarm systems as soon as you move into a new home or when you identify potential security vulnerabilities in your existing space. It's a good idea to install these systems before leaving for an extended period, such as vacations or business trips, to ensure your home is adequately protected while you're away.

DOOR AND WINDOW ALARMS

MATERIALS

- Magnetic sensors, with user's manual (two parts—a magnet and a sensor—will be installed on the door or window frame and the corresponding door or window itself)
- Adhesive or screws
- Alarm unit (if not built into sensor unit), with user's manual
- Batteries
- Mounting brackets (if needed)

TOOLS

- Screwdriver set (flathead and Phillips)
- Drill
- Tape measure
- Level
- Hammer
- Pliers
- Utility knife
- Chisel
- Safety goggles
- Work gloves

KEY ACTIONS

1. Decide where to install the alarms. For doors, place the sensor on the doorframe and the magnet on the door itself. For windows, place the sensor on the window frame and the magnet on the window.

2. Clean the surfaces where you will install the sensors and magnets. Use rubbing alcohol or a similar cleaner to remove dust and grease, ensuring proper adhesion.

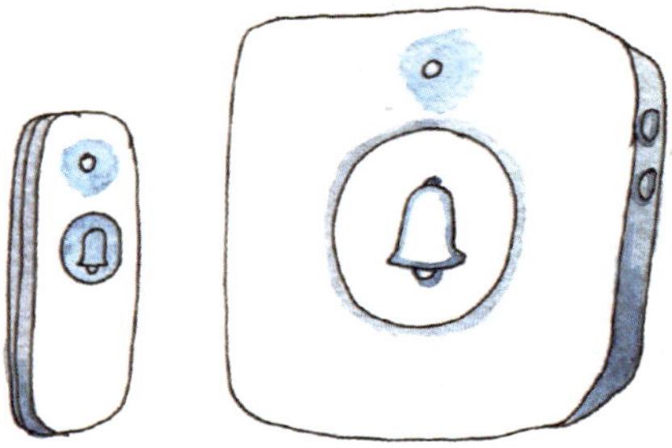

3. If using adhesive, peel off the backing and firmly press the magnet onto the designated area of the door or window. If using screws, mark the holes, drill, and then secure the magnet in place.

4. Repeat the same process for the sensor. Ensure that it is aligned properly with the magnet so that when the door or window is closed, the two pieces are close together.

5. Open the battery compartment of the alarm unit and insert the required batteries, ensuring they are oriented correctly according to the polarity markings.

6. Before finalizing the installation, test the alarm system. Close the door or window to ensure that the sensor and magnet align correctly. Open the door or window to confirm that the alarm sounds.

7. If everything is functioning correctly, secure the alarm in place. Make sure all components are firmly attached to avoid any dislodging during use.

8. Follow the manufacturer's instructions to arm the system. This may involve pressing a button on the unit or using a remote control or keypad.

TIPS

- Regularly check the alarm systems to ensure they are functioning properly. Replace batteries as needed and test the system periodically.
- Mount sensors at a height of 6 to 8 feet to reduce the chances of them being easily disarmed or tampered with by intruders.

- For added deterrence, consider placing fake alarms in visible areas. This can make potential intruders think twice about attempting a break-in.
- If installing outdoors, choose weather-resistant alarms and place them where they won't be exposed to elements like direct sunlight or heavy rain.
- Don't forget about side doors, garage doors, or basement windows. These are often overlooked but can be vulnerable points of entry.

CAUTIONS

- In addition to alarms, ensure that windows have secure locks. Many intruders can bypass alarms if locks are weak.
- Be aware that environmental factors like pets, strong winds, or even heavy rain can trigger false alarms. Choose systems that account for these factors.
- Wireless alarms can experience interference from other electronic devices. Install alarms away from routers, microwaves, and other wireless devices to minimize issues.
- Make sure all household members know how to operate the alarm system, including how to arm and disarm it, as well as what to do in the event of an alarm.
- For smart alarms, ensure you have a stable internet connection. Loss of connectivity can disable remote monitoring and alerts. Consider installing some battery-operated ones as well for backup.
- Relying solely on alarms can create a false sense of security. Combine alarms with other security measures such as cameras, motion lights, and strong locks.

MOTION-ACTIVATED LIGHTS

There are many different types of motion-activated lights. If you want to hook your lights into the grid, I recommend hiring a professional to install your system. If you would like to maintain security and safety in grid failure, consider models that are independent of electricity and run instead on battery or solar power.

MATERIALS

- Motion-activated light fixture (battery-operated or solar-powered models with built-in motion sensors), with user's manual
- Mounting bracket or adhesive strips
- Batteries
- Solar panel (if needed)

TOOLS

- Screwdriver
- Drill (optional)
- Level

KEY ACTIONS

1. Select an area with good visibility and appropriate coverage for motion detection, ideally where the light will not be obstructed.
2. If using screws, mark the positions and drill holes as necessary. If using adhesive, clean the surface to ensure proper adhesion.
3. Attach the light fixture to the wall or surface using brackets or adhesive strips, ensuring it's secure.

4. If battery-operated, insert the required batteries according to the manufacturer's instructions.

5. Activate the light to ensure it detects motion properly and adjust the sensitivity if needed.

TIPS

- Install lights where they can effectively detect motion, such as near entry points, walkways, or driveways.
- Mount the lights at an appropriate height, typically 6 to 8 feet off the ground. This height helps avoid interference from pets while ensuring effective motion detection.
- If installing solar-powered lights, place the solar panel in a location that receives direct sunlight for at least three to four hours to ensure optimal charging.
- Before permanently mounting the fixture, test the light by holding it in place to ensure it detects motion correctly and illuminates the desired area.
- Make sure the motion sensor has a clear line of sight to detect movement effectively.

CAUTIONS

- Avoid placing lights near moving objects (like trees or branches) that could trigger false alarms.
- If using battery-operated lights, choose high-quality batteries for longer life and better performance.
- For larger areas or multiple entry points, consider using more than one motion-activated light to ensure comprehensive coverage.

BASIC CHIME OR BELL SYSTEM

The simplest iteration of this is a trip wire perimeter where a fine wire or microfilament is placed low to the ground and attached to a set of bells or chimes that make a noise if the wire is disturbed. We are going to discuss battery-operated systems, avoiding anything that needs to be attached to an electrical circuit.

MATERIALS

- Battery-operated chime or bell unit, with user's manual
- Batteries
- Mounting hardware

TOOLS

- Screwdriver
- Drill (optional)
- Level

KEY ACTIONS

1. Decide where to install the chime or bell (typically near entryways or areas where you want to alert others).
2. Insert batteries into the chime or bell unit according to the instructions.
3. Use a pencil to mark the desired installation spot for the chime and, if applicable, the push button.
4. If using screws, hold the chime unit against the wall and mark the screw holes. Drill holes if necessary, insert anchors, and secure the unit with screws. If using adhesive strips, clean the surface, apply the strips, and press the chime firmly against the wall.

5. Repeat the mounting process for the push button, ensuring it is easily accessible.
6. Press the push button to activate the chime and ensure it is audible. If there's a motion sensor, check that it activates the chime upon detecting movement.

TIPS

- Select a location for the chime that is easily accessible and can be heard clearly.
- Make sure to clean the chime and push button periodically to prevent dust buildup.

CAUTIONS

- Check the batteries every six months and replace them as needed.

46 CONSTRUCT AN EMERGENCY TOILET

When the grid goes down, many people immediately stress about their phone and then think about their supply of toilet paper. It can be easy to forget that our traditional plumbing systems depend on electricity to pump water into household toilets. A lack of electricity leaves households without a reliable means to dispose of waste. Not having a functional toilet can lead to serious hygiene and health risks, including the spread of diseases and contamination of water sources. With an emergency toilet solution, you can manage waste effectively, ensuring that you and your family can maintain a level of comfort and hygiene during a crisis.

When it comes to building an emergency toilet, you have a few options, depending on the resources available to you and probable duration of the emergency situation. We'll discuss the following three options:

1. The bucket toilet
2. The camping toilet
3. The long drop

Regardless of the option you choose, it's important to follow proper sanitation practices. Always have hand sanitizer, disinfectant wipes, and toilet paper on hand to maintain hygiene after using the emergency toilet. Additionally, plan for waste disposal, ensuring that any bags or waste materials are stored securely and disposed of properly when normal services resume.

WHEN TO DO IT

Depending on the method you choose, these projects are usually set up after an emergency has taken place. However, it's a good idea to ensure you have the materials available and ready to go prior to needing them.

THE BUCKET TOILET

MATERIALS

- 5-gallon bucket or large, sturdy container
- Trash bags (heavy-duty)
- Absorbent material (sawdust, cat litter, shredded newspaper, or peat moss)
- Toilet seat (optional)

KEY ACTIONS

1. Line the inside of the bucket or container with a heavy-duty trash bag. Ensure the bag is large enough to fold over the edges of the bucket for easy removal.
2. Place a layer of absorbent material at the bottom of the lined bucket. This will help absorb moisture and control odors. Use 1 to 2 inches of the material.
3. If using a toilet seat, place it securely on top of the bucket to provide a comfortable seating option. Ensure it is stable and can support weight. Some options for this are a pool noodle, portable toilet seat, or even just a plank of wood.
4. Choose a discreet and private location for the bucket toilet. If using indoors, ensure the area is well-ventilated to help reduce odors.
5. When you need to use the toilet, sit as you would on a regular toilet. After use, cover the waste with more absorbent material to minimize odors.

6. When the trash bag is full, carefully pull it out of the bucket, tie it securely, and store it in a designated area until it can be disposed of properly in line with local waste management guidelines or, if that is not an option, buried in the backyard.

TIPS

- After using the bucket toilet, use hand sanitizer or disinfectant wipes to clean your hands just as you would with a normal toilet. You can also wipe down the toilet seat and surrounding area as needed.
- Have a plan for secure storage of filled trash bags until they can be disposed of properly.
- If using indoors, ensure good ventilation to help reduce odors.
- Ensure the area remains clean and hygienic.

CAUTIONS

- Use disposable gloves for handling waste during cleanup.
- Periodically check the bucket for odors or leaks and replace the trash bag and absorbent material as necessary.

THE CAMPING TOILET

MATERIALS

- Ready-made portable toilet unit (typically includes toilet seat and waste container)
- Specialized waste bags designed for use with portable toilets

KEY ACTIONS

1. Find a suitable and discreet location for your portable camping toilet. Ensure the area is flat, stable, and away from high-traffic zones. If using outdoors, choose a spot that has some privacy.
2. Open your portable camping toilet according to the manufacturer's instructions. If the toilet requires assembly, follow the provided steps to ensure it is securely put together.
3. Place a waste bag inside the waste container of the portable toilet. Make sure the bag fits securely and is properly positioned to catch waste. Some bags may have an absorbent gel or powder that helps control odors.
4. If your portable toilet has locking mechanisms or stabilizing features, engage them to ensure the toilet remains stable during use.
5. When you need to use the toilet, carefully lift the lid and sit down as you would on a regular toilet.
6. After using the toilet, cover the waste with additional absorbent material if required. Dispose of used toilet paper in the waste bag or a separate container, per the manufacturer's recommendations.
7. When the waste bag is full, carefully tie it closed to prevent leaks and odors. Follow the instructions for disposal provided by the manufacturer.

TIPS

- Use disinfectant wipes or spray to clean the toilet seat and surrounding area after each use to maintain hygiene.

CAUTION

- Store the sealed waste bag in a designated area until you can dispose of it in accordance with local waste disposal guidelines.

THE LONG DROP

MATERIALS

- Shovel or trowel
- Piece of wood or a sturdy board (for seat)
- Tarps, wood panels, or a portable privacy tent
- Absorbent material (sawdust, peat moss, or shredded newspaper)
- Cover (optional)

KEY ACTIONS

1. Select a discreet area away from water sources, trails, and high-traffic zones. Ensure the location is relatively flat and stable for safety and comfort.

2. Using a shovel or trowel, dig a pit at least 3 to 4 feet deep and about 2 feet wide. The depth helps contain odors and allows for better decomposition of waste. Make sure to pile the excavated soil nearby for covering the pit later.

3. Use a strong board to create a seat over the pit. This can be a simple flat plank wide enough to sit on securely. Ensure it is stable and positioned over the pit.

4. Set up privacy walls using tarps, wood panels, or a portable privacy tent around the toilet area to provide a discreet space for users. Make sure this shelter is high enough to prevent visibility.
5. Place absorbent material near the toilet area. This will be used to cover waste after each use to minimize odors and promote decomposition.
6. When ready to use the toilet, sit over the pit as you would on a regular toilet. After use, cover the waste with a handful of the absorbent material to control odors.
7. If desired, place a removable cover or lid over the pit when not in use to keep animals out and reduce odors.
8. Once the pit becomes full it will need to be filled in with soil. Use the excavated soil from the original pit to cover the waste and mark the area to avoid future use.

TIPS

- Have a plan for when the pit becomes full, whether it involves burying waste or moving to a new location. If you can, avoid burying toilet paper in the pit, as it may not decompose well. Use a trash bag to dispose of used toilet paper.

CAUTIONS

- Choose a spot away from water sources to prevent contamination. Make sure you fill in the hole before the waste reaches the top. About a handspan is a good amount of cover.

47 MANAGE INSECTS AND PESTS

One of the primary concerns following a disaster—be it a flood, hurricane, or earthquake—is the potential for health risks associated with pests. Many insects and rodents are carriers of diseases that can spread rapidly. For instance, mosquitoes can transmit viruses such as West Nile and Zika, while rodents may carry diseases like hantavirus and leptospirosis. With health-care systems often strained after a disaster, the risk of disease outbreaks can increase dramatically. Thus, effective pest control becomes essential to safeguarding public health and preventing the escalation of these threats.

Food safety is another critical area impacted by pest control. Disasters can disrupt food supplies and storage, leaving food items vulnerable to contamination by pests. Insects and rodents can taint food with their droppings, urine, and body parts, which can lead to foodborne illnesses. Implementing pest control measures is vital to ensuring that any available food remains safe for consumption, thus protecting the health of individuals and communities.

The psychological impact of pests should not be overlooked either. The presence of insects and rodents can exacerbate stress and anxiety for individuals already grappling with the aftermath of a disaster. A clean, pest-free environment contributes positively to the mental health and overall well-being of those affected, fostering a sense of safety and comfort.

WHEN TO DO IT

Prepare for the possibility of pest infestations prior to an emergency situation.

MATERIALS

- Insecticides and pesticides: Chemical or natural insecticides that target specific pests, such as mosquitoes, ants, cockroaches, and rodents
- Rodenticides: Poison bait specifically designed for rodent control; ensure these are used carefully to prevent accidental poisoning of non-target animals or pets

- Traps
- Insect repellents: Sprays or lotions containing DEET, picaridin, or natural alternatives like citronella oil, used to repel insects, particularly mosquitoes and ticks (see page 277)
- Barrier sprays: Residual insecticide sprays that can be applied around the perimeter of properties to create a barrier against invading pests
- Sealants: Caulk or other sealants to close gaps and cracks in buildings where pests may enter; includes sealing around windows, doors, and utility lines
- Boric acid: A natural insecticide effective against ants, cockroaches, and other pests; can be used in bait stations or as a dust application in cracks and crevices
- Waste management supplies (trash bags, bins, and pest-proof compost containers)
- Cleaning supplies
- Gloves, masks, and goggles
- Pest-proof storage containers

KEY ACTIONS

1. After a disaster, evaluate the environment for potential pest problems. Look for standing water, debris, and food sources that could attract pests. Identify the types of pests that may pose a threat in your area.
2. Thoroughly clean the affected area. Remove any debris, standing water, and food sources that may attract pests. Proper waste management is essential; use pest-proof trash bags and bins.
3. Inspect buildings and structures for gaps, cracks, or openings where pests could enter. Use caulk or sealants to close these entry points and reduce the likelihood of infestations.
4. Place traps strategically in areas where you suspect pest activity. For rodents, use snap or live traps. For insects, consider glue traps in corners or near food sources. Ensure traps are placed out of reach of children and pets.

5. If necessary, apply insecticides to areas where pests are prevalent, following the manufacturer's instructions. For rodent control, use rodenticides in tamper-resistant bait stations to ensure safety.
6. Regularly check traps and bait stations to assess pest activity. Replace or reposition traps as needed. Keep an eye on areas where you've applied insecticides or other treatments.
7. Review the effectiveness of your pest control measures. Assess whether additional interventions are necessary.

TIPS

- Store emergency preparedness supplies in pest-proof containers. Periodically check to ensure they have not been infested or contaminated.
- Regularly dispose of waste, clean food storage areas, and eliminate standing water to reduce potential breeding grounds for pests.
- Keep updated on local pest trends and outbreaks, especially after disasters.

CAUTIONS

- Always read and follow the instructions on the labels of pesticides and insecticides. Improper use can lead to health hazards and environmental damage. Pay attention to recommended application rates, safety precautions, and reentry intervals.
- When handling chemicals or traps, wear appropriate protective gear such as gloves, masks, and goggles. This helps prevent skin contact, inhalation of fumes, and eye irritation.
- Store pesticides, rodenticides, and other chemicals out of reach of children and pets. Use childproof containers and secure storage areas to prevent accidental exposure.

- Follow local regulations for disposing of unused or expired pesticides and chemicals. Do not pour them down drains or throw them in regular trash, as this can pose environmental hazards.
- Be aware of any allergies or sensitivities among individuals who may be exposed to pest control products. Consider using natural or less toxic alternatives if allergies are a concern.
- Do not mix different pesticides unless explicitly instructed. Combining chemicals can create dangerous reactions and increase toxicity.
- When applying pesticides indoors, ensure proper ventilation to minimize inhalation of fumes. Open windows and doors to allow fresh air to circulate during and after application.
- If implementing pest control measures, make sure family members know about the methods being used so they can interact with them safely.

NATURAL PEST DETERRENTS

- Certain essential oils, such as peppermint, tea tree, and lavender, can act as natural insect repellents. Mixing a few drops with water in a spray bottle can help to deter pests like mosquitoes and ants.
- Used coffee grounds can deter ants. Sprinkling coffee grounds around entry points can create a barrier that ants are reluctant to cross due to the strong smell.
- A simple mixture of soap and water can be an effective pesticide for soft-bodied insects like aphids and spider mites. Spray the solution directly on the pests to suffocate them.
- Sprinkle food-grade diatomaceous earth in pet areas or carpets to control flea populations. It dehydrates fleas without using chemicals and is safe for pets when used appropriately (follow manufacturer's instructions).
- Certain plants naturally repel pests. For example, marigolds can deter nematodes and aphids, while basil can repel mosquitoes and flies. Incorporate these plants into your gardens.
- Vinegar is an effective cleaning agent that can help eliminate pheromones left by pests. Cleaning surfaces with a vinegar solution can disrupt pest trails and deter them from returning.
- Sprinkling salt around entry points creates an inhospitable environment for slugs and snails while being relatively safe for the surrounding area.

48 RUN EMERGENCY DRILLS

One of the primary benefits of emergency drills is that they familiarize you with the protocols and procedures to follow during a crisis. When people know what to do, where to go, and how to respond, they are more likely to act swiftly and efficiently. For instance, fire drills are commonly practiced in schools and workplaces to ensure that everyone knows the quickest and safest routes to exits. Regularly conducting these drills not only helps individuals memorize evacuation routes but also reinforces the importance of staying calm in chaotic situations.

Another vital aspect of emergency drills is the opportunity to identify and address potential weaknesses in emergency plans. For example, during a tornado drill, participants can assess whether designated safe areas are truly secure and accessible. If a problem arises—such as blocked exits or inadequate communication systems—these issues can be addressed before an actual emergency occurs, thereby enhancing overall safety.

There are several types of emergency drills that you and your family should consider practicing:

- Fire
- Earthquake
- Tornado warning
- Intruder
- Medical emergency

By practicing these drills on a regular basis, you can promote a calm and proactive response to an emergency should it arise.

EXAMPLE OF A FAMILY FIRE DRILL

Step 1: Planning

1. Draw a floor plan of your home and mark at least two exits from every room.
2. Choose a safe meeting point outside, such as a tree, mailbox, or the end of the driveway.

3. Assign roles. For example, one adult helps young children or pets, and an older child pretends to call emergency services once outside.

Step 2: Announce the Drill

1. Tell the family, "This is a practice fire drill."
2. Once everyone is familiar with the process, try unannounced drills to make it more realistic.

Step 3: The Drill

1. Activate your smoke alarm test button to simulate the fire alarm.
2. Everyone moves quickly but calmly to the nearest safe exit.
3. Practice crawling low to the ground as if there is smoke.
4. Teach that hot doors should not be opened. To check, touch the door with the back of your hand first.
5. Go directly to the meeting place outside.

Step 4: Next Actions

1. Once gathered, designate someone to pretend to call emergency services.
2. Remind everyone never to go back inside for belongings.

Step 5: Review

1. Discuss how it went.
    - Did everyone know where to go?
    - Was the meeting point clear?
    - Did anyone get stuck or confused?
2. Record how long it took. The goal is under 2 minutes.

WHEN TO DO IT

Conduct drills regularly, ideally at least once or twice a year, to reinforce preparedness and keep procedures fresh in everyone's minds. Additionally, drills should be scheduled after significant changes to ensure that all individuals are familiar with the latest protocols. See above for a sample drill.

MATERIALS

- Printed copies of the emergency response plans and procedures
- Communication devices
- Timer

KEY ACTIONS

1. Decide which emergency scenario you want to practice, such as a fire, earthquake, or severe weather. Each situation may require different responses and procedures.
2. Create or access your emergency plan for this scenario.
3. Prepare any necessary materials, such as printed copies of your emergency plan, flashlights, first-aid kits, and any props that may help simulate the emergency scenario.
4. Discuss the specific emergency scenario you will be practicing and what actions everyone should take during the drill.
5. Decide on a clear signal to indicate the start of the drill, such as a whistle, bell, or verbal cue. This helps everyone understand when to begin the exercise.
6. Execute the drill according to your emergency plan.
7. Time the drill to see how long it takes and if you improve over time. Set a goal to beat your previous time, as this will foster a sense of urgency and excitement.
8. Once the drill is completed, ensure that everyone gathers at the predetermined meeting point outside the home. This reinforces the importance of knowing where to go after evacuation.
9. After the drill, hold a family meeting to discuss what went well and what could be improved. Encourage open communication and feedback from all family members about their experiences during the drill.

10. Based on the discussion, make any necessary adjustments to your emergency plan or practice areas that need improvement. This could include revising escape routes or identifying additional safety measures.

TIPS

- Use the opportunity to teach family members other important skills, such as basic first aid, how to use a fire extinguisher, or how to contact emergency services.
- Plan to conduct these drills regularly, ideally at least twice a year, to keep everyone familiar with the procedures and ensure the plan remains effective as circumstances change.
- Include role-playing scenarios where family members take on different roles, such as a "lost pet" or "injured family member." This helps practice communication and problem-solving during emergencies.
- Use apps or games related to emergency preparedness to make learning about safety fun. There are various interactive resources that can help your family understand procedures while enjoying technology.
- After conducting a drill, let family members create mock news reports or videos about what happened during the drill. This can be a fun way to reflect on the experience and reinforce learning.
- Establish a reward system where family members earn points or small prizes for participation and completing tasks during drills. This can motivate everyone to take the drills seriously while making it feel rewarding.
- If you have pets, include them in the drills. Teach family members how to safely evacuate with pets and practice carrying them if necessary. This ensures that everyone is prepared, including furry family members.

CAUTIONS

- While it's beneficial to create realistic drills, avoid scenarios that may be too frightening or overwhelming, especially for younger children. Focus on preparedness without inciting panic.
- Always prioritize safety during drills. Clear any obstacles that could cause injury during evacuations and ensure that all participants understand the importance of following safety protocols.
- Be mindful of each family member's comfort and anxiety levels. Some individuals may be more sensitive to emergency scenarios. Encourage participation without forcing anyone to engage in a way that makes them uncomfortable.
- Demonstrate calmness and positivity during drills. Your demeanor can greatly influence how family members perceive the exercise. If you appear anxious, it may increase their stress levels.
- Conduct drills in a controlled environment with minimal distractions. Turn off phones or televisions to help participants focus on the drill and the instructions being given.
- While practicing, ensure that the scenarios align with realistic situations your family might face. Avoid overly complicated or unlikely scenarios that could confuse participants.
- Ensure that instructions are clear and concise. Miscommunication can lead to confusion during a drill, undermining the practice's effectiveness.
- Keep your emergency plan updated and review it regularly. Changes in family dynamics, living situations, or emergency protocols may necessitate updates to the plan.

49 COMPILE PREPAREDNESS CHECKLISTS

Preparedness checklists are essential tools that provide a clear, organized framework for outlining necessary actions, gathering resources, and establishing plans for effective crisis response. By systematically listing essential tasks, these checklists reduce the likelihood of overlooking critical steps, ensuring that you are well equipped to handle various scenarios, from natural disasters to medical emergencies. This proactive approach not only fosters confidence but also enhances overall readiness, allowing you to respond more effectively in high-stress situations.

Additionally, preparedness checklists promote better communication and coordination among family members during crises. They serve as a reference point during training and drills, helping everyone understand their roles and responsibilities, which is crucial for effective response. Furthermore, they encourage you to assess potential risks and take preventive measures, contributing to risk mitigation and overall safety. Ultimately, the psychological reassurance that comes from having a well-thought-out plan in place alleviates anxiety, making preparedness checklists invaluable for building resilience and ensuring a more successful outcome in times of crisis.

Preparedness checklists that you should consider having include:

- Emergency kit checklist
- Evacuation plan checklist
- Medical and health checklist
- Pet emergency checklist
- Community resources and services checklist
- Seasonal preparedness checklist

I've provided samples of each on the following pages for inspiration. Feel free to add and subtract items in a way that makes the plan relevant to your home and your situation. You may need more of something because you have a lot of kids or less of something due to allergies or dietary requirements.

EMERGENCY KIT CHECKLIST

Food & Water Supplies

- ☐ Water (1 gallon per person per day)
- ☐ Nonperishable food (3 meals per person plus snacks per day)
- ☐ Water filter or purifying tablets
- ☐ Water container
- ☐ Manual can opener
- ☐ Small stove and 2 gas canisters

Tools & Equipment

- ☐ Flashlight (either rechargeable or with spare batteries)
- ☐ Fire starter
- ☐ Extra batteries
- ☐ Multi-tool or fixed-blade knife
- ☐ Whistle
- ☐ Cell phone with chargers
- ☐ Battery-powered or hand-crank emergency radio (NOAA Weather Radio recommended)

First Aid & Health

- ☐ Home first-aid kit (see page 214)
- ☐ Prescription medications
- ☐ Emergency blanket

Hygiene & Sanitation

- ☐ Toilet paper
- ☐ Personal hygiene items
- ☐ Hand sanitizer

Documents & Essentials

- ☐ Cash (small denominations)
- ☐ Important documents (identification, insurance)
- ☐ USB flash drive with scanned documents

Pet Supplies (if applicable)

- ☐ Pet food
- ☐ Water
- ☐ Medications (if needed)
- ☐ Leash
- ☐ Crate

Clothing & Shelter

- ☐ Warm clothing
- ☐ Sturdy shoes
- ☐ Sleeping bags or blankets

EVACUATION PLAN CHECKLIST

- ☐ Identify evacuation routes.
- ☐ Designate meeting points.
- ☐ Confirm communication plan (see page 235).
- ☐ Grab your emergency kit (see page 284) or go bag (see page 227).
- ☐ Make arrangements for pets.
- ☐ Inform relatives and/or neighbors of plans.
- ☐ Secure important documents.
- ☐ Ensure access to transportation.
- ☐ Stay informed about local alerts.
- ☐ Preload maps and alerts apps (e.g., FEMA, Red Cross, local emergency services).

MEDICAL AND HEALTH CHECKLIST

- ☐ Prescription medications
- ☐ Over-the-counter medications
- ☐ Home first-aid kit (see page 214)
- ☐ Medical equipment (glucose monitor, inhaler, etc.)
- ☐ Health insurance information
- ☐ List of allergies and medical conditions
- ☐ Emergency contact information for health-care providers
- ☐ Copies of medical records
- ☐ Vaccination records
- ☐ Personal hygiene items
- ☐ Thermometer
- ☐ Emergency medical information card
- ☐ Extra pair of glasses or contact lenses
- ☐ Sunscreen and insect repellent
- ☐ Special dietary needs supplies (if applicable)

PET EMERGENCY CHECKLIST

- ☐ Pet food
- ☐ Water and water bowls
- ☐ Leash and collar
- ☐ Pet carrier or crate
- ☐ Identification tags
- ☐ Medical records and vaccination history
- ☐ Medications
- ☐ First-aid kit for pets
- ☐ Comfort items (blanket, toy)
- ☐ Waste disposal bags
- ☐ Extra leash or harness
- ☐ Emergency contact information for veterinarian
- ☐ Pet-grooming supplies
- ☐ Travel documents (if applicable)
- ☐ Pet safety information (behavioral notes, special needs)
- ☐ List of pet-friendly shelters or hotels

COMMUNITY RESOURCES AND SERVICES CHECKLIST

- ☐ Local emergency services (fire, police, medical)
- ☐ Emergency shelters
- ☐ Food banks and pantries
- ☐ Community health clinics
- ☐ Red Cross or similar organizations
- ☐ Local government emergency management office
- ☐ Disaster relief organizations
- ☐ Mental health services
- ☐ Transportation services for evacuees
- ☐ Community centers or gathering places
- ☐ Volunteer organizations
- ☐ Local news and radio stations for updates
- ☐ Utility companies (for outages and emergencies)
- ☐ Animal shelters or rescue organizations

SEASONAL PREPAREDNESS CHECKLIST

Winter

- ☐ Rock salt or ice melt
- ☐ Snow shovels
- ☐ Winter clothing (coats, gloves, hats, shoes)
- ☐ Blankets and sleeping bags
- ☐ Extra food supplies (nonperishable)
- ☐ Emergency heating source (space heater, blankets)
- ☐ Winter tire chains or snow tires
- ☐ Firewood or alternative heating fuel
- ☐ Vehicle emergency kit (jumper cables, flares)
- ☐ Ice scraper and snow brush for vehicles
- ☐ Insulated water storage or bottles

Summer

- ☐ Sunscreen
- ☐ Insect repellent
- ☐ Hydration supplies (water bottles, electrolyte drinks)
- ☐ Lightweight, breathable clothing
- ☐ Emergency cooling supplies (fans, cooling towels)
- ☐ Fire extinguisher
- ☐ Outdoor safety gear (hats, sunglasses)

50 ORGANIZE A MASTER PROJECT PLAN

Having a project planner schedule is essential for effective time management and organization, ensuring that all aspects of a project are carefully planned and executed. It provides a clear road map, outlining tasks, deadlines, and responsibilities, which helps to keep family members accountable and focused. By visualizing the project timeline, potential bottlenecks and challenges can be identified early on, allowing for proactive problem-solving. Additionally, a project planner fosters communication among family members, facilitating collaboration and ensuring everyone is aligned with the project's goals. Ultimately, utilizing a project planner schedule enhances productivity, minimizes stress, and increases the likelihood of successful project completion.

Each project in this book includes guidance on the optimal timing to start it. To begin your project plan, use the sample priority list (see page 289) to prioritize the projects that interest you most.

Once you've identified your priorities, break each project down into manageable steps (see page 290). Assign each step to specific family members, along with clear deadlines. Setting target dates and assigning accountability partners significantly increases the likelihood of completing the overall project successfully.

Add your own flair to the templates provided to make them more interesting to you and your family. That way you'll be more likely to succeed. Including rewards such as a trip to the park or a gold star system for tasks completed can also make it more fun.

Ultimately, a master plan boosts productivity, reduces stress, and greatly increases the chances of successfully completing your projects.

SAMPLE PRIORITY LIST

High Priority Tasks	Date Due

Medium Priority Tasks	Date Due

Low Priority Tasks	Date Due

Additional Tasks	Date Due

SAMPLE PROJECT PLANNER

Objective

Timeline

Notes

Milestones

Tasks	Assigned To	Deadline

SAMPLE EMERGENCY PLAN

Our Household Address

Members of our Household	
Name:	Phone #:
Name:	Phone #:
Name:	Phone #:
Name:	Phone #:
Name:	Phone #:
Name:	Phone #:
Name:	Phone #:

Alternate Meeting Place	Pet Plan

Contacts	
Family Doctor:	Phone #:
School:	Phone #:
School:	Phone #:
Neighbor:	Phone #:
Neighbor:	Phone #:

FIRST AID

No emergency preparedness book would be complete without a section on first aid. Keep in mind that reading about how to do something is no substitute for hands-on experience in a course taught by professionals. This section is for emergencies or to jog your memory in the heat of the moment.

Recommended procedures are constantly changing as research finds more effective ways of administering first aid, so it is good to update your CPR qualifications every year and your first aid every three.

I have over thirty years of first-aid and wilderness first-aid course experience in three different countries. Qualifications, techniques, and recommended procedures are always evolving—I have used this knowledge often in both urban and wilderness environments, making patients as comfortable as possible and ensuring they were in the best condition they could be once professional medical care arrived.

Your initial response to the scene of an accident will be the same regardless of whether you are in an urban or wilderness setting and if you have brought a first-aid kit or not. That is the DRS ABC of being a first responder.

DRS ABC

This is a handy acronym to help you remember your order of priorities in a first-aid scenario.

D - DANGER

The last thing that you want in an emergency scenario is to put a healthy person in danger. Once you also become a victim, chances of survival diminish. As soon as you see a person in trouble or if you come across someone who looks injured, check the area for danger. What was the mechanism of injury? Is it something that will cause injury to you? Has the danger passed, or can you safely remove the victim from danger?

If you have seen someone fall to the ground because a branch broke, it should be safe for you to move in to help them immediately. If someone has crashed into power lines, make sure they aren't active or compromising the scene before moving in. Consider waiting for professionals to arrive.

Always take a moment to assess the scene. Acting proactively will usually have a better result than acting impulsively. This can be seen with the amount of people who drown trying to rescue loved ones from riptides in the ocean. If they took a moment to grab a floatation device before entering the water, their success would almost be guaranteed. Diving in without one ends up putting two lives in danger.

R - RESPONSE

Once you have established that there is no danger to either party, it is time to assess their level of consciousness. If the person is yelling or screaming, their trachea (the body's largest airway) is clear, and they have a pulse. Move on to your secondary assessment.

If the person is still, try to get a response from them. Don't shake them violently because you are still unsure what else may be wrong with them. Simply touch their shoulder and ask some questions. Give them a moment to respond. If they moan, open their eyes, or answer a question, you can move on to your secondary assessment.

If there is no response, assume they are unconscious and roll them into the recovery position (see below). If the mechanism of injury was a fall from a height or a head hit, assume that the spine may be damaged and take extra care when moving into the recovery position to keep the spine as straight as possible.

THE RECOVERY POSITION

An unconscious person is placed in this position because it allows the airways to stay open and fluids to drain out if need be. The tongue muscle may fall back and block the airways of an unconscious person on their back. To move someone into the recovery position, lie them on their back, bend the arm nearest to you so that their hand cushions their head, and bend the opposite knee to you. Pull the person over onto their side so that their knee holds them on their side. Tilt their head back slightly to ensure their airways are open.

S - SEND FOR HELP

This was a more recent addition to first-aid teachings because it was a step that people in urban environments kept forgetting to do. If you are with a group of people and have access to cell phone service, designate someone to call 911. If you are by yourself, take a second to dial, put the phone on speaker, and continue with your assessment.

A - AIRWAY

Look for any obstructions in the mouth. Check and scoop out anything that you see inside their mouth that shouldn't be there. If their breathing is wheezy or noisy, there may be something blocking their airway. Do not shove your fingers too far down their throat to try to clear it out, as this will trigger the gag reflex. If you can't see anything, check their breathing.

B - BREATHING

If the airway is clear, check for breathing by listening, looking, and feeling. Place the back of your hand near their mouth to see if you can feel air moving by. Place your hand high on their chest to see if you can feel a rise and fall. Listen to see if you can hear their breath moving in and out.

If the person is breathing, keep them on their side and continue to do your secondary assessment. If the person isn't breathing, turn them onto their back and commence CPR.

C - COMPRESSIONS

There are many schools of thought as to how many compressions per minute and how many rescue breaths, but as long as you are doing compressions and breaths so that blood is continuing to circulate through the body and oxygen can get to the brain, you are doing more than nothing. Thirty compressions to two breaths is considered fairly standard. In times of airborne pathogens, choosing whether to do rescue breaths will be up to you. If you know the person and know their health status, you may make a choice to do assisted breathing. If the patient is a stranger, it is okay to just give chest compressions, knowing that some air will be moving in and out of their lungs while you are massaging their heart.

HOW TO DO CHEST COMPRESSIONS:

1. Kneel by the person's side.
2. Place the heel of one hand in the center of the person's chest.
3. Place the heel of your other hand on top of the first hand.
4. Interlock your fingers.

5. Position yourself vertically above the person's chest.
6. Keep your arms straight and press down on the sternum about three inches.
7. Keep a pace of approximately one hundred compressions per minute.

HOW TO DO ASSISTED BREATHING:

1. Pinch the soft part of the patient's nose closed.
2. Tilt their head back slightly.
3. Use a pinch grip to raise up their chin.
4. Take a normal breath and place your lips around the person's lips, creating a seal.
5. Blow steadily into their mouth while watching to see if the chest rises.
6. Remove your mouth and let the chest fall.
7. Take another normal breath and blow into the patient's mouth.

If the chest isn't rising when you breathe into their mouth, try tilting the head back a little more or look for an obstruction in their airway.

You will need to continue CPR until the person is breathing on their own, a medical professional has arrived to take over, or your health and safety are compromised. You may be able to swap in and out with other people if you aren't alone, allowing for some rest between rounds.

If you are in an urban environment, there is a good chance that a defibrillator is somewhere nearby and someone has managed to access it for you. Using one of these would be the next step if this is available and they haven't responded to the first few rounds of CPR.

A person will either recover and start breathing or they won't. It is important to understand that chances of resuscitation are slim without intervention from either a defibrillator or a professional medical team. Do not feel like you have failed if you are unable to resuscitate someone. The failure would have been in not possessing the knowledge to at least try.

SECONDARY ASSESSMENT

If the person is conscious, or unconscious but breathing, you will need to check to see what else can be done to keep them alive. This is where the scenarios become too numerous for me to list. There are situations that need to be dealt with immediately or the person will deteriorate, such as deadly bleeds and venomous snake bites. Most other scenarios will allow you some time to call for help or administer basic first aid; however, with deadly bleeds, seconds and minutes count, so I will touch on them now.

DEADLY BLEEDS

A severed artery can lead to someone bleeding out in minutes. I have lost count of the times I have seen people cutting something with a knife while resting the object on their leg. Fatigue and lapses in concentration can have disastrous consequences when combined with sharp objects.

Deadly bleeds are wounds where the blood is pumping from the wound, the bleeding doesn't slow down or stop with pressure, and/or the blood is quickly soaking through the bandage. Deadly bleeds can be internal or external. External ones are more obvious, as there will be blood soaking clothing or flowing from the wound. Internal bleeds can generally be observed as swelling or bruising at an impact site. If the swelling or bruising is in the abdominal region, call for help and evacuate the person as soon as possible.

For all deadly bleeds, immediate evacuation and medical assistance will be required, so calling or sending for help is a priority.

If the object is still in the body, do not attempt to remove it. If it is protruding from the wound, immobilize it in place so it doesn't move around and create further damage internally. Removing projectiles could open up arteries and the patient can bleed out in minutes.

The first thing to do with a massive bleed is to apply pressure to the wound. Grab some absorbent material if available or use your hands to try to stem the bleeding. In some cases, wounds will stop bleeding when pressure is applied to them. Elevate the wound above the level of the heart if possible. If you have absorbent dressing and a bandage available, place it on the wound and wrap in place with a bandage. You want a firm bandage but not a tourniquet. If the wound bleeds through the dressing, do not remove the initial dressing; simply apply another layer and another bandage.

STRAIGHT-SIDED DEEP CUT

If the wound is a slice cut, hold the two sides of the wound together as you apply pressure. Once bleeding has slowed, flush out the wound, check the depth of the wound, and assess whether it is likely to continue to bleed. If the wound requires stitches, use butterfly stitches from your first-aid kit to hold the sides of the wound together and cover with clean, nonadhesive dressing.

WOUNDED LIMB

If the wound is deep and to a limb, apply pressure, elevate the area above the heart, and cover with a pressure bandage, making sure that it is firm but not tight enough to cut off blood flow to the extremities.

ABDOMINAL WOUND

If the wound is in the abdomen, and there are no internal organs visible, treat as above. If there are internal organs visible or protruding, **DO NOT TRY TO STUFF THEM BACK IN THE CAVITY**. Cover the wound with a moist, sterile dressing and bandage firmly. Make the patient comfortable but do not give them anything to eat or drink.

CHEST WOUND

If the wound is in the chest and a wheezing noise can be heard through the wound, assume that a lung has been punctured. Cover the hole with a dressing that can create a pressure seal. This can be a ziplock bag or something similar. Tape around three edges of the square, allowing one side to remain free. Immobilize until help can arrive.

TOURNIQUET

Unless a limb has been completely shattered or torn off or the femoral artery (the main one in your leg) has been compromised, try to avoid applying a tourniquet. A tourniquet is a last resort as it will probably result in the loss of a limb once applied in the field. If your choice is to lose a life or lose a limb, apply a tourniquet and do not take it off. A tourniquet is a bandage that restricts all blood flow to a body part. Write the time and date that you applied it. The only person who should remove a tourniquet is a medical professional once the person has made it to a medical facility.

Once you have dealt with DRS ABC and deadly bleeds, you should be able to take your time with further assessment.

Remember that this is a very rough and quick guide to basic first aid. Please take the time to complete a professional course with a registered provider. You never know when your knowledge could make a difference.

CONCLUSION

As we wrap up this exploration of home safety and resilience, take a moment to appreciate the knowledge you've gained. With fifty engaging projects covering your basic needs and more, you now possess a wealth of tools and insights to create a secure environment for you and your loved ones should a disaster occur.

But remember, preparedness isn't just a checklist; it's a journey of learning and growth. Each project you tackle, no matter how small, contributes to your peace of mind.

This book serves as your first step into a world where you can take control of your safety and well-being. Embrace the process and make it a family affair! Involving your children, partners, or friends not only makes the tasks more enjoyable but also instills essential survival skills in the next generation. As you work together, you'll create lasting memories and a stronger bond, all while building a safer home environment.

The best gift you can give yourself and those you care about is peace of mind. Knowing that you are prepared for whatever life may throw your way is empowering. It allows you to live fully in the present without the weight of uncertainty looming overhead.

As you continue on this path, remember that even small efforts can have a significant impact. Celebrate your achievements and be open to revisiting and refining your projects as your understanding and needs evolve. Here's to a future filled with security and resilience—one enjoyable project at a time.

INDEX

ABOUT THE AUTHOR

KY FURNEAUX is one of the world's most recognized survival experts and the author of nine survival and preparedness books, including *Survive*, *When the Grid Fails*, *Adventure Awaits*, and *Surviving the First 36 Hours*.

A former Hollywood stunt performer and Taurus World Stunt Award winner, she now dedicates her career to teaching practical survival skills that help people thrive in both wilderness and everyday life.

Ky served as a survival consultant on the blockbuster film *Send Help* and has been featured as a survival expert across numerous international television networks. She is best known to audiences as a fan favorite on Discovery's hit series *Naked and Afraid*, and for her own short series *Outback Lockdown*.

Through her books, media work, and leadership programs, Ky helps people develop the mindset, resilience, and practical skills needed to face challenges with confidence—using survival principles not just for the outdoors, but for navigating life itself.

ABOUT CIDER MILL PRESS BOOK PUBLISHERS

Cider Mill Press publishes exceptional books that combine creativity and craftsmanship. As an imprint of HarperCollins Focus, we specialize in premium cookbooks, cocktail and spirits guides, and illustrated gift books, all distinguished by compelling content, striking design, and a commitment to quality in every detail. Cider Mill Press sets the standard for books that inform, inspire, and elevate everyday moments. Learn more at cidermillpress.com.

CIDER MILL PRESS
BOOK PUBLISHERS

"Where Good Books Are Ready for Press"
501 Nelson Place
Nashville, Tennessee 37214 USA